Proceedings of the
Fourth Annual Tri-Service Conference
on the
Biological Effects of Microwave Radiation

Proceedings of the
Fourth Annual Tri-Service Conference
on the

Biological Effects of Microwave Radiation

Volume 1

16–18 August 1960
New York University Medical Center

COLONEL GEORGE M. KNAUF
Conference Chairman

Sponsored by
Rome Air Development Center
Air Research and Development Command
United States Air Force
Griffiss Air Force Base, New York

HERBERT S. BROWNSTEIN
Air Force Project Monitor

Edited by
MARY FOUSE PEYTON
Department of Industrial Medicine
New York University Post-Graduate Medical School

Springer Science+Business Media, LLC

1961

Library of Congress Catalog Card No. 61-11807
© 1961 Springer Science+Business Media New York
Originally published by Plenum Press, Inc. in 1961.
Softcover reprint of the hardcover 1st edition 1961
All rights reserved

ISBN 978-1-4899-5629-3   ISBN 978-1-4899-5627-9 (eBook)
DOI 10.1007/978-1-4899-5627-9

CONTENTS

Welcome   1

> NORTON NELSON
> New York University

Radio Frequency Environment   3

> OLIVER G. TALLMAN
> Rome Air Development Center

Chairman's Remarks   9

> GEORGE M. KNAUF
> Patrick Air Force Base

Basic Problems in Measuring RF Field Strength   15

> HARRY R. MEAHL
> General Electric Company

**Microwave Instrumentation for the Measurement of
Biological Effects**   23

> JOSEPH H. VOGELMAN
> Capehart Corporation

**Generation and Detection of Pulsed X-Rays
from Microwave Sources**   33

> ANTHONY P. DE MINCO
> Rome Air Development Center

Quick Formulas for Radar Safe Distances   47

> H. S. OVERMAN
> US Naval Weapons Laboratory

Some Engineering Aspects of Microwave Radiation Hazards   55

> R. S. ENGELBRECHT
> W. W. MUMFORD
> Bell Telephone Laboratories

**Development of a Garment for Protection of Personnel
Working in High-Power RF Environments**   71

> MARTIN R. REYNOLDS
> Filtron Company, Inc.

The Time Constants of Pearl-Chain Formation   85
M. SAITO
H. P. SCHWAN
University of Pennsylvania

The Effect of Microwave Radiation (24,000 mc)
on the Male Endocrine System of the Rat   99
SAMUEL A. GUNN
THELMA CLARK GOULD
W. A. D. ANDERSON
University of Miami

Effects of Radio-Frequency Energy on Human Gamma Globulin   117
SVEN A. BACH
ANTHONY J. LUZZIO
ARNOLD S. BROWNELL
US Army Medical Research Laboratory

Longevity and Cellular Studies with Microwaves   135
SUSAN PRAUSNITZ
CHARLES SÜSSKIND
PAUL O. VOGELHUT
University of California

Phantom Experiments with Microwaves at the
University of Rochester   143
HERBERT MERMAGEN
University of Rochester

Relative Microwave Absorption Cross Sections of
Biological Significance   153
A. ANNE
M. SAITO
O. M. SALATI
H. P. SCHWAN
University of Pennsylvania

Biological Effects of Microwave Energy at 200 mc   177
C. H. ADDINGTON
C. OSBORN
G. SWARTZ
F. P. FISCHER
R. A. NEUBAUER
Y. T. SARKEES
University of Buffalo

Effects of 2450 mc Microwaves in Dogs,
    Rats, and Larvae of the Common Fruit Fly   187
        GORDON W. SEARLE
        ROGER W. DAHLEN
        CHARLES J. IMIG
        CHARLES C. WUNDER
        JOHN D. THOMSON
        JOHN A. THOMAS
        WILLIAM J. MORESSI
        State University of Iowa

The Effect of 2450 mc Radiation on the Development
    of the Chick Embryo   201
        CLAIRE VAN UMMERSEN
        Tufts University

Specific Thermal Effects of High Frequency Fields   221
        VICTOR T. TOMBERG
        Biophysical Research Laboratory

Microwave Radiation in Relation to Biological Systems
    and Neural Activity   229
        J. FLEMING, JR.
        L. PINNEO
        R. BAUS, JR.
        R. MCAFEE
        Tulane University

Neurological Effect of 3 cm Microwave Irradiation   251
        ROBERT D. MCAFEE
        CAROLYN BERGER
        PHILIP PIZZOLATO
        New Orleans Veterans Administration Hospital
        Tulane University

Biomedical Aspects of Microwave Irradiation of Mammals   261
        JOE W. HOWLAND
        RODERICK A. E. THOMSON
        SOL M. MICHAELSON
        University of Rochester

Changes in the Ascorbic Acid Content in Lenses of
    Rabbit Eyes Exposed to Microwave Radiation   285
        LORENZO O. MEROLA
        JIN H. KINOSHITA
        Harvard University
        Massachusetts Eye and Ear Infirmary

**Preliminary Results of Studies of the Lenticular Effects of
Microwaves Among Exposed Personnel** 293

MILTON M. ZARET
MERRIL EISENBUD
New York University

A Review of Unanswered Biological Hazard Operational Problems 309

JOHN E. BOYSEN
Wright-Patterson Air Force Base

**Similarities and Differences between the Technical Aspects of
the Navy HERO Program for Ordnance and the
Personnel Hazard Program** 319

JAMES N. PAYNE
US Naval Weapons Laboratory

Appendix 327

Index 329

Welcome

NORTON NELSON
Department of Industrial Medicine
New York University Medical Center
New York, New York

IT IS A VERY GREAT PLEASURE to welcome you here on the occasion of the Fourth Tri-Service Microwave Conference. I speak both for New York University Medical Center and for its component, the Institute of Industrial Medicine, in assuring you of our appreciation of the honor shown us in holding the meeting here.

The Institute of Industrial Medicine has as one of its major interests the conduct and encouragement of research aimed at control of environmental hazards, and it is thus doubly gratifying to see this assembly here for purposes so closely akin to our own.

The ultimate objective of this meeting and of continuing studies will be to define how microwaves can be used safely and without harm. Although this objective is clear and unambiguous, the means required are often involved and intricate as is illustrated by the wide spread of talents and interests present here at this conference. I might say that it has been my general observation that raising this question in a forthright manner is never really a practical impediment to the technical operation in question; on the contrary, it generally turns out that the groundless fears and timidity arising from ignorance are a far greater interference than a realistic control program based on understanding.

Study of hazard from microwaves is still in a relatively early phase. It is at this stage that one asks the question, is there a risk, and if so, whence does it come, what are its characteristics, and what is the extent and nature of the physiological effect? Beyond the interest and urgency arising from practical and humanitarian considerations, this is an exceedingly interesting scientific quest, and, as a

1

glance at the program will show, one that encompasses nearly the whole span of modern scientific and technical knowledge. On the present program are papers that range from fundamental considera- tion of the molecular responses of living tissue to microwaves, through reports concerned with the physical characteristics and measurement of microwave energy, to the organizational and administrative problems in establishing programs of protection.

The conference now about to open is not likely to resolve all of your problems. It may, and in fact probably will, raise new ones; perhaps one of the most useful fruits of such a gathering as this is the opportunity to discover the weaknesses of current methods and secure suggestions for an improved attack on the problem.

I look forward to a productive and interesting session.

Radio Frequency Environment

OLIVER G. TALLMAN
Rome Air Development Center
Rome, New York

THE EMPHASIS at the Rome Air Development Center in the past year has been on obtaining better understanding of the noted biological effects from some of our prior work, but there has been no substantial amount of deviation from the hazard purposes of our program. While we continue to recognize that the basis for this entire program is that of the so-called personnel hazard, many signs have evidenced themselves in the continuance of this research program indicating that we should explore other uses of electromagnetic energy as it may affect biological specimens. If the services can someday in the near future get into a position to help fund these other interesting things which we have noticed from time to time in our research over the last 6 years, we see great promise in this type of work. Although we have now worked in this field for several years, we can not claim to know it all. It would seem, nevertheless, that the direction of our research has been correct and profitable in that we have continued to maintain an adequate protection environment for all personnel concerned with Air Force operations.

RADC not only co-ordinates and leads this program, but conducts the Air Force contractual program of research from Rome and also provides the field instrumentation developments to be employed in the field in the measuring of this energy. I should refer back to my statement that we have continued to provide adequate protection, and add to that. While our safety procedures are considered adequate, an unfortunate accident involving X-radiation exposure from an Air Force radar took place during this past year. To us it has served to emphasize a need for better training of all personnel associated with

3

4   **Oliver G. Tallman**

this field, and improved dissemination of information and perhaps a new emphasis on automatic alarm equipment. And again, more effort is needed in the foolproofing of our equipment and in the procedures employed in its use.

In each of our three prior annual conferences, we have had the benefit of Air Force personnel reviewing the basics of the physics phenomena associated with our problems with particular emphasis on the future equipments which may become problems. And to a lesser degree, we have heard some predictions of higher and higher powers. The powers of future radiating equipments may in the next 10 years be beyond our wildest dreams of 4 years ago. This is due largely to man's jump-off into space. I'm speaking of man's interest and not necessarily in the literal sense.

While this energy will be employed in space operations where antennas should be expected to point skywards, I think we may look forward also to leakages in the immediate ground areas and these will serve to keep us busy in ensuring the continued safety to personnel involved in Air Force operations. It is appropriate to briefly review our current or our standard operational situations on the ground dealing with potential rf hazards. These, after all, are the Air Force basis for this work. But we will specifically exclude X-radiation, which will be discussed later in this program in some detail by Mr. De Minco also of Rome Air Development Center.

Figure 1 is a re-drawn graphic portrayal of an actual Air Force Base in which, by the time the radar gets to be sited, the buildings are already in. The little white rectangles indicate existing buildings

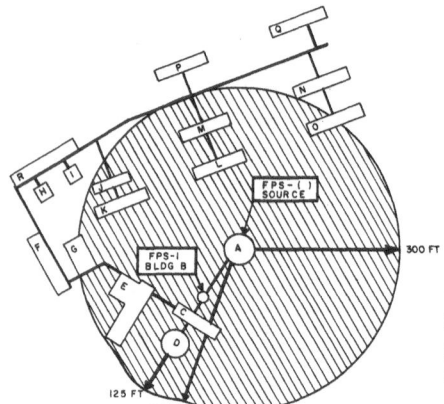

Fig. 1. The shaded area represents a power density of .01 w/cm² or greater, using the near-field correction.

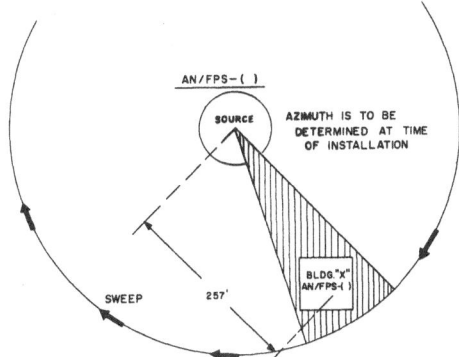

Fig. 2. Sector blanking to prevent radiation hazard.

on an Air Force Base. The center of the circle A is the location of a radar. The area of A is simply an extrapolation of the calculated power density of .01 w/cm² which we have set as the tolerance level for our Air Force operations. In the case of circle D we have another radar set which, when the antennas are pointing simultaneously in one direction, enlarge the 300-ft circle to a distortion of 375 ft in that one direction. Now again, one may wonder about the location of the buildings within the so-called danger zone. While this can be done on a pre-determined and a calculated basis, it is nevertheless proven out by the use of field intensity measurements at the time of installation to ensure that the calculated powers which appear to be of negligible attenuation through the walls of a building, and are not so negligible for attenuation through concrete walls, and are even better in the case of screened windows, are in fact safe. I could have also drawn in air strips adjacent to it, particularly the ramps in front of hangars, the maintenance hangars, the ramps for re-fueling operations, and the special weapons storage which are incident to many Air Force Bases. But this is the area of the immediate problem, so this serves to orient us in an Air Force type of operation.

In Figure 2, we have taken a slightly different look at this subject. Here we have a critical area: a building which must be there and must operate something and, therefore, it has personnel associated with it. In this case we can use the technique of sector-blanking in which the radio frequency is automatically shut off as it sweeps around and thus excludes irradiation of this building or any personnel in that particular area. If you happened to have a particular sector where you need coverage in this area, you would have to re-locate the radar rather than the building so as to move off this axis.

In Figure 3 we are concerned with height above ground. This building itself is approximately 80 ft high with the center point of the antenna 125 ft above earth. At a distance of something over 100 ft we have .02 w/cm², exceeding our standard. At a slightly increased distance we have the .01 w/cm² which is our standard. On the axis itself, if we run all the way out at almost 100 ft elevation, we have .01 w/cm² at a distance of 300 ft. Fortunately, buildings of this height are not found in an Air Force operational situation, and this gives us some degree of freedom for ground operations in the immediate vicinity of the antenna. We make very careful leakage, and backscatter, and near-field measurements in this area to see what the powers are in actuality.

In Figure 4 we see a somewhat different type of antenna, one radiating more power and one which may be expected to receive increasing use in the Air Force. This antenna is on a cliff, with the ocean waves as a good rf mirror. In this case the beam is so shaped that the .01 w/cm² point extends 12,000 ft forward of the antenna to a height of approximately 2500 ft and a width of 6000 ft. This

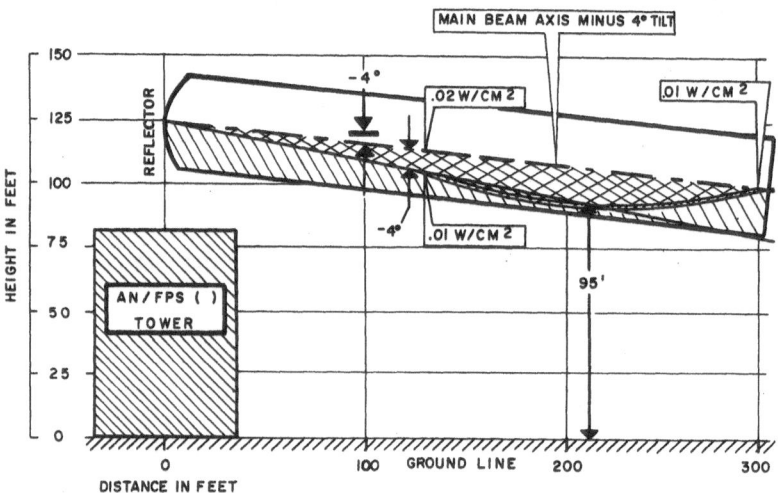

FIG. 3. Power computed off the main axis of the beam for .01 w/cm² using the near-field correction. $(P_T G_T / 4 \pi d^2) F_f = P_0$.

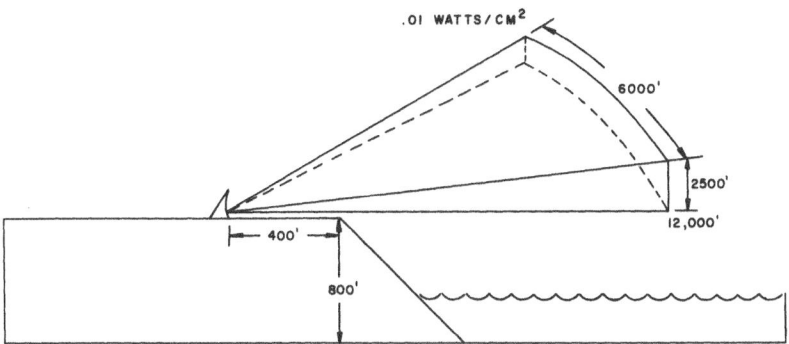

Fɪɢ. 4. Three-dimensional power density distribution for high-power scanner.

gives us problems in the cliff area in regard to leakages, scatters, reflections, etc.

You will hear mention later in the program of rf shielding suits. These will assist in this type of operation where conditions necessitate the presence of personnel.

Chairman's Remarks

GEORGE M. KNAUF
Patrick Air Force Base, Florida

IT IS INDEED A PLEASURE to see so many old friends here. It is truly heartwarming to know that so many of you are still with us in this search for radar hazards. I extend to you my personal thanks and the thanks of the Tri-Service Committee for your support.

As all of you know, there have been in the past many trials and tribulations connected with the management of our research program. I can report to you today, however, that we appear to be back on the track again and should be able to look forward to a period of much smoother sailing. I am sure you join me when I express my pleasure in noting that our effort is back in the hands of our good friends at the Rome Air Development Center. I am sure you all agree that their support and help have been vital to any measure of success we may have enjoyed in this effort.

It is appropriate that I extend a special "thank you" to Mr. Tallman who has given so generously of his time and talents to this work. He has been a stalwart supporter and a most patient counsellor.

I could name many more who have given me invaluable help over these past five or six years. Time just does not permit. It is sufficient that I say that I come before you deeply grateful for all the assistance I have received.

There are a number of things relative to our work that I think you should be made aware of. First, the current status of our MSEL of .01 w/cm². I am indeed pleased to say that up to today we have not seen any research data which shake our faith in the validity of this arbitrary safe exposure level which we sponsored some five years ago. As many of you know, this level has been challenged on occasion. It has been suggested that perhaps we selected this level because it was an operationally acceptable level and not because it offered the promise of a safe environment for the worker. It is true that this level has proven to be an operationally feasible one. However, could you have heard the protests of our operational colleagues when they first were told to live with this level, I am sure you would have concluded that operational suitability was not the basis for selecting .01 w/cm² as a MSEL. As time has gone by, it has been impressed upon the equipment operators that they can function effectively and still adhere to this level. At the same time our research has not demonstrated a need to modify this level to avoid injury to personnel in ordinary work situations. In certain selected situations additional protection is indicated such as the wearing of protective clothing. You will hear more about these isolated problem areas as this meeting progresses. You should bear in mind, however, that these situations are exceptions and in no wise the rule.

If any modification of our .01 w/cm² MSEL looms on the horizon, it is in the direction of being able to perhaps liberalize the allowable safe exposure criteria in the case of those frequencies found to be in the main reflected from the human body and not absorbed in any appreciable quantity.

Now about the acceptance of this level. As you know, there is virtually no disagreement in this country on this level as ensuring a safe working environment. It is true that in some cases individual users of microwave energy have developed their own interpretation of the application of this level to their peculiar working environment, but in general they do not disagree with the basic safe level.

I think you might be interested to know something about the acceptance our work has received abroad. I have just returned from a visit to SHAPE, NATO, and the United Kingdom. SHAPE and NATO, which represent some fourteen European countries, have adopted and published our standard. All of the reports of your research work are to be found in the reference libraries of these agencies. In the United Kingdom the picture is even more cheerful. Here I found that not only did they accept our level, but they have developed a keen and healthy interest in advancing the progress of our studies by way of certain applied engineering support of their own. In the course of my London visit I had an opportunity to talk to the Russian delegation to the 3rd International Congress on Medical Electronics. These folks surprised me to some degree by telling me that they too have endorsed our .01 w/cm². They further informed me that prominent in their reference library on the subject of radar hazards were translations of the proceedings of the past three annual conferences we have held on this subject. I learned that they too are engaged in some studies in this field and that their results generally agree with ours. The USSR Academy of Medical Sciences has suggested an interchange of papers on this subject which I hope to be able to accomplish. It seems appropriate that our two groups of biologically oriented people exchange ideas. Such cross fertilization may well be productive of extremely valuable data.

Now to the business at hand. I would like first to put your minds at ease on a subject which I am sure must bother you all. You have no doubt noted that I am scheduled to function as the chairman of these proceedings for the entire three day meeting. Let me assure you that I am fully aware of the misgivings such news must bring to you. I assure you that I plan to offer you some relief by looking around for some good friend who will take over this job for one or more of these sessions. You may now relax, secure in the knowledge that relief is not too far away.

Some of you may remember our last meeting when we were so graciously provided for on the campus of the University of California in the hands of our friend and colleague Dr. Süsskind. If you remember this meeting at all, you will also no doubt remember the difficulty I had in trying to keep the program on schedule. In spite of my best efforts and using all measures short of mayhem I was remarkably unsuccessful in this respect. This time it will be different. I have been empowered by due authority to resort to any weapons necessary to terminate any or all presentations at the expiration of the time allocated for such papers. I hope last summer's offenders will take note.

Seriously, we have put together an ambitious program and one which I am sure will prove to be our best. This last can only be true if all of you assist me in keeping our schedule. There is little point in prolonging papers to a point of pain at a meeting of this sort. It seems to me that the approach here should be to swiftly and concisely highlight the material for the audience and rest on the published paper to supply the technical and burdensome detail. I do sincerely hope you will assist me in making this our best conference yet.

Now a word about the program itself. In the past we have asked each investigating group to present a review of the work they had accomplished in the year gone by. Unfortunately, research progress does not always proceed in such a manner as to make such reports vital and interesting. Altogether too frequently we found that one or another of our investigators was faced with the problem of presenting a paper at an inopportune time in the conduct of his research task. This led inescapably to the inclusion in our program of some fairly nonproductive periods. This time we have changed our approach. We first asked each investigating group whether they had any material which they felt should be brought to your attention at this time. If their project had developed such material, we of course encouraged them to present such data to this group. If they had nothing new to report, they were just as energetically encouraged to remain silent.

At the same time we contacted all of the individuals and agencies known to us to be interested in this problem of microwave effects, and extended to them an invitation to present to this audience papers on subjects felt to be of general interest or wide application.

The result was most gratifying. By this approach we found ourselves in a position to put together what promises to be a most beneficial and productive program. In the past our practice of presenting research reports has operated to deny program time to many individuals in allied fields of interest who might have presented extremely valuable and stimulating material to this audience.

As I study this year's program, I am impressed with the wide span of interests covered. We will, of course, be brought up to date on significant progress in various of our research projects. This is essential if the purpose of this meeting is to be served. But in addition to this updating of our so-called "Report to the Stockholders," we propose to present material in which I am sure you are all vitally interested but which does not fall in the realm of biological research. We will be given a look at what the future promises in the way of new equipment. We will be brought up to date on the progress being made in the development of protective clothing to be made available to personnel whose duties may require that they enter areas where the ambient microwave power density exceed our MSEL.

Instrumentation problems will be discussed and some new ideas in that area presented.

We also will have reviewed for us the X-ray problem as it relates to the operation of certain of our higher power tubes. This is an extremely important and a very timely subject. I am afraid that in the face of our very determined effort to instruct our people on the subject of rf hazards we have allowed them to drop their guard in relation to this second and vitally important problem of X-ray exposure. A recent unfortunate accident has again emphasized the need to keep this area of personnel risk in the foreground in our programs of personnel instruction.

We have included in our agenda this year, toward the end of the program, a talk which I hope will be of value to all of us in providing some measure of guidance in planning the conduct of our studies during the coming year.

As many of you know, in the Air Force the Air Materiel Command has responsibility for the industrial medical program that is carried out at all of our Air Force installations. They are charged with translating data produced by groups such as this into useful and workable guidance for our people in the field. In the Air Force they might well be considered our biggest and best customer. Of course they are at the same time the agency which receives the questions from the field and often are the first to sense a new problem area in our operational spectrum. It is our desire and purpose to develop data which will be of use to using commands such as the Air Materiel Command in coping with field problems. Believing their interests in this area to be typical in most respects of the interests of similar agencies in our sister services, we have asked that the Air Materiel Command present to this group a review of the problem areas which still harass them, with emphasis on those areas in which they feel vital data relative to biological hazards are still lacking. We are hopeful that such a first hand presentation of the current parameters of the hazards problem will prove valuable in ensuring that our efforts during the year ahead of us will be directed along productive lines.

In all, I sincerely believe this year's program offers rich promise of being our best yet. Again, I am delighted to see that so many of you are still with us in our effort to solve the problem of radar hazards.

Basic Problems in Measuring
RF Field Strength

Harry R. Meahl
General Engineering Laboratory
General Electric Company
Schenectady, New York

INTRODUCTION

IN THE PAST, emphasis has been put on far-field solutions to antenna patterns and on measurement of a weighted average of the field strength at a chosen height above ground and along radials drawn from the transmitter using tuned loop or tuned dipole antennas. All this was proper, since the object of analysis and measurement was to predict and confirm the service area of a transmitter.

Some radar and communications transmitters have been built having such high average power output that it has become necessary to predict and confirm the values of the near field of the antennas in order to prevent the exposure of personnel to field strengths in excess of 0.01 w/cm², the present ceiling for continuous exposure. The emphasis must be upon seeking the maximum values of irregular intensifications of field strength which occur in the near field at particular points in space. This philosophy of seeking the maximum instead of an average is narrowed further to consideration of areas of the order of 1 cm² instead of 1 m² or more by the fact that it is necessary to protect a part of one's body, such as the eye, rather than the whole body.

The value of field strength, 0.01 w/cm², mentioned above is the rf exposure level established by the U.S.A.F. in accordance with Technical Order No. 31-1-511A of October 1957. It is believed to be low enough so that personnel may be exposed to it continuously without having it affect health. By agreement, it is applied as a ceiling without regard to frequency of the transmitter or duration of exposure. This broad application is made necessary by the limited knowledge of the biological effects of exposure to rf fields. There are at least seven investigations being made in universities and colleges to accumulate knowledge on the effects of rf fields, other than heating, so one may look forward to the establishment of different exposure levels for different ranges of frequency in the future.

Basic Problems

The basic problems in measuring rf field strength for rf hazard surveys are:

1. A new philosophy of measurement must be used; i.e., seek the maxima in small volumes of space expecting concentrations which are not predicted by calculations.

2. Most field strength measuring instruments, even those advertised as field instruments for hazard use, are affected by immersion in fields approaching the ceiling exposure level and must be modified. Some are subject to permanent damage from temporary overloads which are often encountered in field strength surveys.

3. The measurement team must be protected from exposure to rf fields in excess of the exposure level while measuring them.

4. Observations must be made at a single frequency and over a broad band of frequencies at the same time.

5. Two or more antennas having widely different characteristics must be compared to discover the effects of multiple reflections. For example, comparing a small loop with a dipole often reveals spotty concentrations which would be unsuspected from a survey using a dipole alone.

Solutions — Locating Maxima

The difficult problem of measuring the strong concentrations of rf field at a radar site is simplified by using indicators to find the maxima and then choosing the antennas to use with the power measuring equipment. The General Electric Type NE-2 neon lamp used in a resonant loop, as shown in Figure 1, has been useful. An observer may explore a large area quickly by whirling one of these on a cord. He can see the variations in field strength above a certain level in three dimensions and consequently can locate maxima.

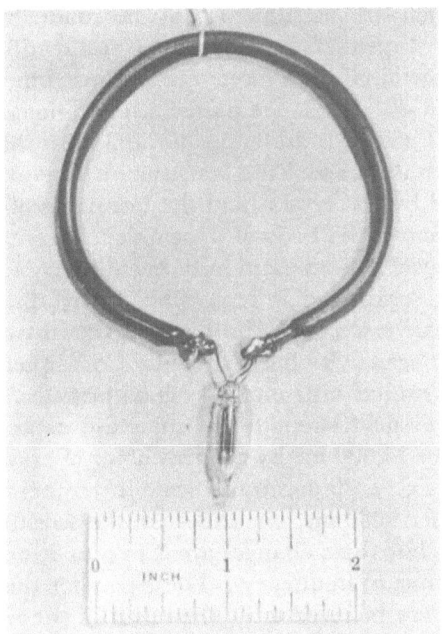

FIG. 1. General Electric Type NE-2 neon lamp
in a resonant loop.

Experience has shown that for frequencies up to 1000 mc these neon lamps in resonant loops have lighted at field strengths less than the exposure level. Each one should be calibrated just before it is used and several should be available on each survey, because although the lamps will recover immediately from extreme overloads, the loop may be detuned by squeezing or bending it slightly and thus lower the sensitivity.

At frequencies above 1000 mc the neon lamps may be used with the leads bent into the form of a dipole or cut off near the glass seal. Under these conditions it is likely that field strengths above the exposure level will be required to light them. They are still useful, but must be manipulated and observed from a position free of strong fields. They are also useful if worn on an observer's shirt, where they can warn him when he is exposed to a strong field.

While observers are searching for maxima with neon lamps in resonant loops, which respond to the operating frequency alone, they should also have an indicator which responds to a broad band of frequencies. Such an instrument may be made by connecting a General Electric Type G7C germanium crystal rectifier or its equivalent across the terminals of a direct current instrument, as shown in Figure 2. When the signals are pulsed, an additional capacitance of from 0.25 to 1.0 μfarad is added in parallel with that shown. It is mounted next to the case of the instrument to avoid a shielding of the loop formed by the crystal and the terminals of the instrument which would occur if the body of the capacitor were near the terminals. These must be used cautiously because crystal rectifiers will not survive overloads and it is possible to overload one without causing a full scale deflection. In this case, the instrument often still responds, but its sensitivity has changed. Consequently, the availability of several similar instruments is good practice. These may be made sensitive to field strength less than the exposure level at all frequencies up to 15,000 mc by proper choice of crystal rectifier and dc instrument, e.g., a 20 μamp full scale instrument with a 1N23B crystal rectifier for 9000 mc. As expected, the sensitivity will change with frequency, but these changes are 2 to 5 to 1, not 10 to 50 to 1, over a 2 to 1 range of frequency. These smaller than might be expected changes are believed to be the result of the relatively high rf losses in the loop formed by the crystal rectifier and the terminals of the instrument, which tend to smooth the frequency characteristic.

FIG. 2.   General Electric Type G7C germanium crystal rectifier
and microammeter.

Fig. 3. Tuned resistive loop antenna, 5 cm².

These instruments may be held in the observer's hand and moved about with a twisting, tilting motion when searching for maxima, or they may be mounted on a wooden, plastic, or bamboo rod. Even though these instruments change the rf field one immerses them in, sharp concentrations of field strength can be located quickly and then measured by means of a suitably modified power measuring equipment and two antennas, e.g., a tuned resistive loop and a dipole. A tuned resistive loop is shown in Figure 3. It is made of 4 rf resistors, whose series resistance totals approximately 51 ohms, and a small ceramic capacitor of such capacitance as to tune the loop made by the capacitor and the resistors to the operating frequency of the radar being surveyed. It was made resistive to avoid large changes in impedance with small changes in operating frequency. The loop was tuned to provide a low V.S.W.R. load at the operating frequency for the coaxial cable connecting the loop to the power measuring equipment. The end result is an antenna which has an effective area equal to its geometric area within 30% over a range of frequencies of 2 to 1. The losses in this type of antenna are calculable when the loss in the cable connecting it to the power measuring equipment is known. The power will be equally divided between the antenna and the cable-power measuring equipment combination. The rf field strength is the total power, loss in the antenna plus loss in the cable

plus measured power, divided by the area of the antenna, i.e., milliwatts per square centimeter. When a dipole is used, it is often necessary to add attenuation between it and the power measuring equipment because its effective area is large, $0.13\lambda^2$, at frequencies below
1000 mc.

Agreement will be found between small loop antennas, 1 to
3 cm^2, and dipoles or horns both in the far field and in a near field
resulting from many sources and many reflections, such as may be
found in a scanner building. In the latter case, the orientation of a
dipole will have no effect on its output. However, field patterns may
be formed which cause a dipole to give a low reading. In one case,
near a coaxial line double slug tuner, a dipole antenna delivered a
maximum of 1 mw, but a loop of 1 cm^2 area delivered a maximum
of 2 mw when used to search the same volume. The effective area
of the dipole was 2900 cm^2, so it should have delivered several watts.
In any field pattern which may be described as spotty, a dipole or
horn antenna will provide an average which is naturally lower than
the maximum in the volume being observed. Comparisons of the
field strength observed with a dipole or horn with that observed with
a small loop reveal additional information on the field strength
pattern in space.

Test of Measuring Equipment

Power measuring equipment may be tested by locating it in a
position where field strength greater than the exposure level is expected without connecting an antenna, e.g., in front of a feed horn.
This should be done while the transmitters are off, of course, and the
equipment may then be observed from a position of low field strength
after the transmitters are started. It may be necessary to use field
glasses or a telescope to read the instrument.

Improvement of Shielding

If large deflections are observed in the test described above, the
power measuring equipment should be examined closely to determine how the rf energy is being introduced into the circuit. It may
be found that there are holes or slots in the case, that a power lead
for use in charging the internal battery is acting as an antenna, that
the indicating instrument has a plastic case and scale plate and no

shielding behind it, or that an insulating gasket is used between the panel and the case.

It is good practice to make only one modification at a time and retest to determine its effect.

The use of absorbing-type shielding is often effective. A pad of steel wool sealed inside a thin-walled plastic bag and located inside a case covering a suspected opening may be enough to reduce the deflection to a tolerable size.

However, it may be necessary to add solid copper or copper screen in addition to the absorbing pad or pads.

In the case of a power lead, one may either modify the equipment to make the cord detachable or pack absorbing pads around the folded cord and fasten the assembly to the case with a strong adhesive tape.

Protection of the Measurement Team

Good planning and continuing vigilance are required to protect members of the measurement team from exposure to rf fields above the continuous exposure level while they are finding and measuring maxima of field strength which occur on a radar or communication site.

Whenever possible, the first survey should be made while feeding the system with a low power source, a 10-w oscillator, for example. This procedure allows the measurement team to locate concentrations of rf field by using sensitive detectors, such as crystal rectifiers in tuned loop antennas or even conventional field strength meters, without risk. The information obtained at low power will guide the measurement team in placing indicators in strategic spots about the site to show where the continuous exposure level is exceeded when the transmitter or transmitters are operating.

When rf field strength in excess of the continuous exposure level is expected or discovered, steps should be taken to obtain the needed data without exposing the personnel. In some cases, it will be possible to build temporary shelters using rf absorbing materials, either alone or in combination with metallic shields. For example, a wall of rf absorbing material may be erected in front of a scanner and a measuring team may obtain data by mounting indicating instruments or antennas on rods or poles made of plastic, dry wood, or dry bamboo. The area which personnel will occupy should be monitored by remote viewing of a number of different indicators before personnel enter. Of course, a number of indicators should be watched while personnel are present, and radio contact should be maintained with the control room to make it possible to remedy any condition of excessive field strength quickly by having the scanner moved or the power removed.

Each member of the measurement team may wear several of the NE-2 neon lamps on his person, some in resonant loops when the operating frequency is below 1000 mc and some without leads. These give a visual warning that a certain level has been exceeded, and a power measuring equipment may then be used to determine the field strength. As experience with particular neon lamps is gained during a survey, reasonable estimates can be made from their brightness. Then the operator can make measurements at those locations indicated to have the greatest field strength while he keeps himself at positions of tolerable field strength.

Remote viewing and manipulating will be used to an increasing degree as the power outputs are increased.

Conclusions

The basic problems in measuring rf field strength in the near field are difficult, but not insoluble. No single instrument provides all the needed information, but proper use of a number of different instruments does.

Microwave Instrumentation for the Measurement of Biological Effects

JOSEPH H. VOGELMAN
Capehart Corporation
Richmond Hill, New York

INTRODUCTION

THE MEASUREMENT of biological effects from microwave energy involves two instrumentation areas: the biological and the electrical. This paper will devote itself exclusively to the electrical problems involved in creating ambient field densities of known value and relating these densities to the measured effects. Consideration will be given to problems of antennas operating in the near field, conditions for beam formation, reflection, and interaction insofar as it affects the production of known ambient field densities. Techniques for direct coupling of microwave energy to selected organs of a biological specimen will be considered. Since cages, containers, and other devices are essential for securing the specimen, the effects of their presence in the microwave field will be analyzed. Specific problems resulting from pulse operation will be considered in terms of the interpretation of the related field and specific problems in peak versus average power measurement. The measurement of the temperature of the specimen, while under electromagnetic radiation, will be analyzed. Specific sources of error will be pointed out with recommendations for minimizing them.

Field Density

The instrumentation associated with the measurement of the biological effect of microwave energy is generally of insufficient power output to permit accurate measurements at sufficiently high field densities in the far field of the antenna. The range at which a particular field density can be achieved for a specific combination of total transmitted power and antenna gain is given by the relationship:

$$R = 5.26 \times 10^{-4} f \sqrt{P_T A/D} \qquad (1)$$

where R is the distance from the antenna in feet, f is the transmitter frequency in mc/sec, P_T is the transmitter average power in watts, A is the cross-sectional area of the antenna aperture in ft^2, and D is the desired field density in mw/cm^2. The equation assumes an efficiency of 50% for the antenna, which is normal for most apertures. Using this relationship, it can be seen that with available power levels and small apertures the range must be small. The near field extends to a distance given by

$$R_N = 2d^2/\lambda = 0.0279 fA \text{ feet} \qquad (2)$$

To ensure operation in the far field, these two equations can be combined to obtain the criteria for the required power for any prescribed aperture and field density. This is given by

$$P_T > 53.1 AD \qquad (3)$$

where P_T is the power output in watts, A is the aperture in ft^2, and D is the required field density in mw/cm^2. From eq. (3) it can be seen that an aperture of 1 ft^2 would result in a density of 100 mw/cm^2 only if the transmitter power exceeded 5310 w. This relationship further indicates the necessity for minimizing the aperture consistent with the frequency being used if far-field measurements are to be made. Table I should serve for the selection of combination of aperture and

TABLE I

Power Requirements for Far-Field Operation

Aperture, ft^2	Required power (in mw/cm^2) for indicated density				
	50	100	200	500	1000
0.1	265.5	531	1062	2655	5310
0.2	531	1062	2124	5310	10620
0.5	1328	2655	5310	13280	26550
1.0	2655	5310	10620	26550	
2.0	5310	10620	21240		
5.0	13280	26550			
10.0	26550				

TABLE II

Minimum Aperture as a Function of Frequency

Frequency, mc/sec	Aperture, ft^2
200	8.1
500	3.24
1000	1.62
2000	0.81
5000	0.32
10000	0.162
20000	0.081
50000	0.032

field densities commensurate with the available power. Table II gives the minimum aperture size suitable for operation as a function of the operating frequency. Since most investigators will not have sufficient power to ensure far-field measurement and exposure of the biological specimen, it becomes necessary to bring the specimen within the near field.

In the near field two phenomena are present which contribute to errors in the determination of the actual field density to which the biological specimen is exposed and, in turn, introduces a questionable factor in the quantitative values for observed effects. The first effect is the cyclic variation in field density in the near field as one proceeds from the antenna aperture outward to the end of the near-field region. This axial power density variation is illustrated in Figure 1 for the specific case of uniform illumination of the antenna aperture. As can be seen from Figure 1, the exact position of the specimen with respect to the aperture will determine the ambient field density. Since the introduction of the specimen into the field moves the cyclic variation, it is almost impossible to predict the actual ambience within the near field for any but the simplest states suitable for exact or good approximate computation. At the same time, measurements of the field density in the near field may not suffice since the field variations are displaced to a different degree by the measuring instrument and the specimen. Where the measuring instrument may indicate a peak, the introduction of the specimen at the same point may result in a minimum of ambient field density.

The second phenomenon which results from the introduction of a biological specimen in the near field of an antenna is an interaction

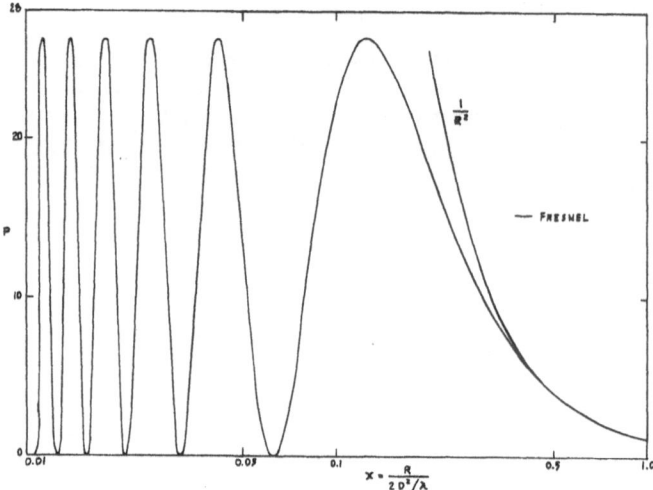

FIG. 1. Axial power density, uniform illumination.

between the specimen and the antenna. This results in an impedance mismatch as seen by the generator of the microwave power. This mismatch may result in a marked change in the power generated as well as in a change in oscillator frequency. These effects depend on the degree of sensitivity of the generator to standing waves and the proximity of the biological specimen to the antenna aperture. Accordingly, near-field measurements of biological effects lack quantitative accuracy since the prediction of the ambient field density is both difficult and inaccurate.

Two approaches lend themselves to more accurate application. The first approach, intended for whole-body radiation of biological specimens, is illustrated in Figure 2. A nonresonant metallic chamber of good conducting material, such as copper screening of 20 by 20 mesh, is used to house the specimen to be radiated. This chamber is connected directly to the transmission line coupled to the transmitter signal source. A bi-directional coupler is incorporated in the transmission line to provide indication of the incident power as well as the reflected power. When the chamber is empty, the highly conductive walls reflect the radio frequency energy with the result that the reflected power is essentially identical with the incident power. Where required, a Faraday Rotation Isolator must be inserted between the signal source and the bi-directional coupler to ensure that the reflected energy does not cause breakdown in the signal source. This type of device should be set up to provide a load for the reflected power away from the signal transmitter tube itself. When the specimen is inserted into the chamber, the difference between the incident power and the relative reflected power is the energy absorbed in the specimen. If the incident and reflected power outputs from the directional coupler are recorded, the animal can be free to wander about the chamber and a record of instantaneous exposure is still available. The most important precaution is that the chamber be nonresonant at the frequency of operation. A nonsymmetrical chamber lends itself most readily to the creation of a nonresonant structure.

For the specific radiation of a single appendage or area of a biological specimen, the instrumentation shown in Figure 3 is recommended. The output of the waveguide structure is terminated either in an expanded section or in an extremely thin-walled iris with an opening the size of the area to be exposed to microwave energy. Spring fingers are used around the waveguide structure to ensure good coupling to the specimen and at the same time minimize leakage radiation. A set of tuners are included in the waveguide structure. The specimen is inserted into the spring fingers so that the enlarged section of waveguide or the iris is in contact with the appendage or the portion of the body to be irradiated. The tuners are adjusted so that the reflected power is reduced as close to

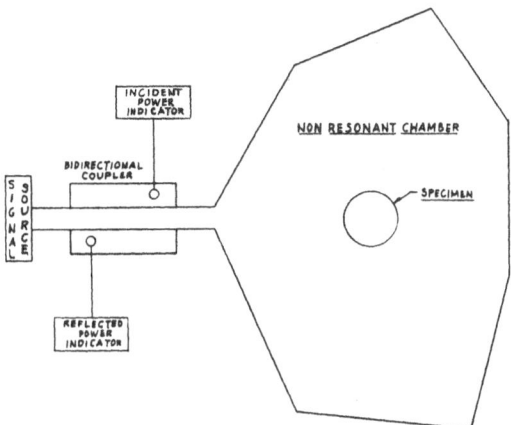

Fig. 2. Instrumentation for whole-body radiation.

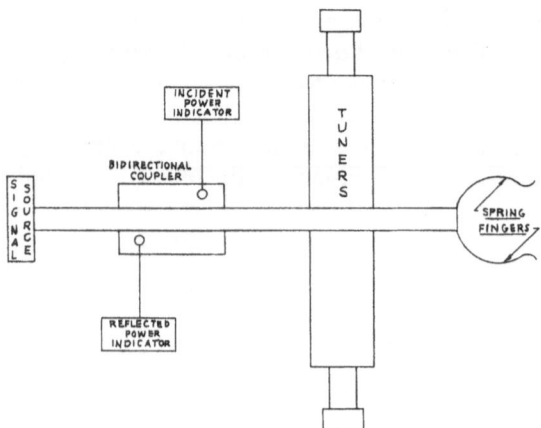

Fig. 3. Instrumentation for radiation of appendage.

zero as possible. The difference between the incident power and the reflected power is the energy coupled to the biological specimen.

The foregoing techniques, designed to be used in lieu of near-field measurements, should result in reproducible data both within the operation of a single experimenter and between experimenters. It should also permit scaling in frequency so that frequency dependence can be determined from the resulting data.

Cages, Containers, and Other Securing Devices

A few basic rules must be observed in the use of cages, containers, or other devices for securing the biological specimens. These units must be made of dielectric materials with as low a loss as possible. Polystyron, polyvinyl chloride, or Teflon are suitable materials for use in this application. Second, the spacing between components of dielectric (the open spaces) must be at least one wavelength in the direction perpendicular to the polarization of the antenna or wave-guide. Third, the use of water coolant must be so confined as to provide an unobstructed path between the waveguide and the specimen, at least one wavelength in diameter. Otherwise, the water will shield the specimen completely from the microwave radiation. Fourth, where absorbing material is used to shield portions of the specimen, the absorbent must be between the exposed area and the source of microwave energy. Since the absorbent converts microwave energy into heat, extending the exposed portion through the absorbent will subject it to both direct thermal heating as well as microwave heating and becloud the meaning of the resultant data. An extremely useful shield where only a limited portion of the specimen is to be exposed is a thin copper sheet with an opening the size of the area to be exposed. This sheet will reflect energy from all portions to be shielded. To ensure that the reflected energy is not returned directly to the signal source, the copper surfaces should form an angle with the perpendicular running between the specimen and the feed aperture (i.e., the axis of the beam of the antenna).

Temperature Measurement

The use of thermocouples or thermistors for the measurement of temperature rise in the specimens, while under electromagnetic radiation, is strongly affected by the energy coupled directly into the thermistor from the microwave field. The dimensions of the specimen and the location of the thermal probe in the specimen determine the degree of coupling of the radio frequency field. When this happens, the measured temperature gradients are no longer a valid indication of the microwave energy converted into thermal energy within the specimen. This can be effectively demonstrated by measuring the resistance change of the thermistor in the radio frequency field if it is suspended in air and comparing it to the resistance change measured when the thermistor is imbedded in various sizes of dielectric balls of extremely low-loss material such as polyethylene or Teflon. The dielectric material forms a matching structure between the thermistor and the radio frequency wave resulting in an increase in the absorption of radio frequency energy by the thermistor. If any heating in the dielectric results, it is usually from contact with the thermistor rather than the reverse. The same phenomenon occurs where the thermocouple or thermistor thermometers are imbedded close to the exposed surface in biological specimens.

A liquid thermometer electrical transducer is recommended instead as a measurement probe to be used in the presence of radio frequency field. The structure consists of a temperature-sensitive liquid separated from a temperature-stable liquid by a small metallic plug in a thin glass tube. The plug forms a sliding disc within the glass tube and is free to move as the temperature-sensitive liquid expands or contracts. The insensitive liquid, which must be compressible, is compressed as the temperature rises. The position of the metallic plug is electrically coupled to external electrodes and the measured output is an indication of the temperature of the specimen. If the temperature-sensitive liquid is a good dielectric of low loss, the heating resulting from the direct radiation will be negligible.

Pulse Microwave Energy

It is essential at this time to introduce a time factor into the specific conditions for biological effect in a microwave radiating field. The validity of the assumption that only the average power is significant does not conform to the basic conditions associated with thermal energy alone. The simple experiment of passing a finger rapidly through a hot flame without burning and the possibility of repeating this experiment at short intervals without cumulative effects should indicate the significance of the time factor. In order to establish more realistic criteria for the effects of time and exposure, as well as peak-to-average power ratio, the following techniques of measurement are proposed. In the first case, the average power-time product would be maintained constant (i.e., the total energy absorbed is a constant). As exposure time is decreased, the power density is increased to maintain the constant energy absorption. The data to be furnished for conducting these experiments are obtained from S-band measurements (approximately 2700 mc/sec) but can be readily scaled to other frequencies. A square aperture of 10.63 in. on each side, tapered to a standard S-band waveguide, would be used in lieu

TABLE III
Power—Time Data

Average power density, mw/cm^2	Time of exposure	
	Minutes	Seconds
165	45	0
660	11	15
2640	2	49
10560	0	42
42240	0	10.5
168960	0	2.6

TABLE IV

Power Exposure Data*

Peak power, w/cm^2	Average power, mw/cm^2	Pulse repetition frequency
500	165	320 pulses/sec
2060	165	80 pulses/sec
8250	165	20 pulses/sec
33000	165	5 pulses/sec
94200	165	1¼ pulses/sec
376800	165	19 pulses/min

*Average power and time are constant. Exposure time, 45 min. Pulse width, 1 μsec.

of very short distances to achieve the required field densities for the very short exposures. A dielectric flange would be used to provide the pressure on the exposed animal during the long-range exposures, while the antenna aperture output flange could be used during the short-time exposures with appropriate reduction in power. Information on power level and exposure time are shown as Table III. The second phase would consist of determining the peak-to-average power ratio effect. In this case, the pulse repetition frequency would be adjusted to compensate for the increase or decrease of heat power so as to maintain a constant average power. The exposure time would be identical at 45 min for all cases. As the peak power is increased, the average power would be maintained by adjusting the pulse repetition frequency. The values to be used are given in Table IV.

Conclusion

Experimental instrumentation considerations have been presented which are intended to result in more reproducible and more quantitative measurements of biological effects of microwave energy. It is strongly recommended that the series of pulse experiments described be conducted. Particularly sensitive areas of the specimen used should be selected to enhance any apparent results. The above types of measurements are considered a critical necessity to the establishment of more realistic criteria for the new high-powered equipment as well as to set the course for future experimentation.

Generation and Detection of Pulsed X-Rays from Microwave Sources

Anthony P. De Minco
Electronic Engineering Branch
Development Engineering Laboratory
Directorate of Engineering
Rome Air Development Center
Rome, New York

INTRODUCTION

Present and future high-power electronic equipment is found to be capable of radiating two potentially hazardous radiations in the electromagnetic spectrum. One is the rf output we are designing the equipment for and endeavor to transmit, and the other is X-radiation which is an unwanted by-product.

In his quest for higher and higher output power from his microwave generators, man has almost outdone himself. In 20 years he has come from average outputs of 10 w to present outputs of almost 1,000,000 w. In addition to the consideration of microwave radiation hazards with increased power, we must seriously consider the potential hazard of X-radiation. In order to produce increased powers, bigger and better high-power tubes are required. This necessitates an increase in plate voltages and plate currents which provide the necessary ingredients for more efficient penetration and higher ionizing intensities. Until recently, the X-radiation intensities from Air Force electronic equipment were not considered as serious potential hazards as the rf or microwave radiation hazard.

Many of our microwave generators contain components whose X-radiation outputs far exceed any current or industrial X-ray generators. As yet we have not found a use for this unwanted by-product, and its presence is evident only by the manifestation of a strong potential hazard. We have learned to live with hazards in other areas, providing we recognize, evaluate, and institute means for personnel protection. This paper will be concerned in general with generation and detection of pulsed X-radiation emanating from high-power microwave generators.

Generation of X-rays

A brief review of the basic concepts of X-ray production or generation is offered so as to orient the reader to the nature of this particular potential hazard. X-rays are a form of radiant energy, like visible light. Figure 1 shows their location in respect to the rest of the electromagnetic frequency spectrum. Their distinguishing feature is the extremely short wavelength—only about 1/10,000 of the wavelength of visible light or even less. It is this characteristic that is responsible for the X-rays' ability to penetrate materials which absorb or reflect ordinary light. X-rays are produced when electrons, traveling at high speed, collide with matter. In the usual type of X-ray tube, an incandescent filament supplies the electrons,

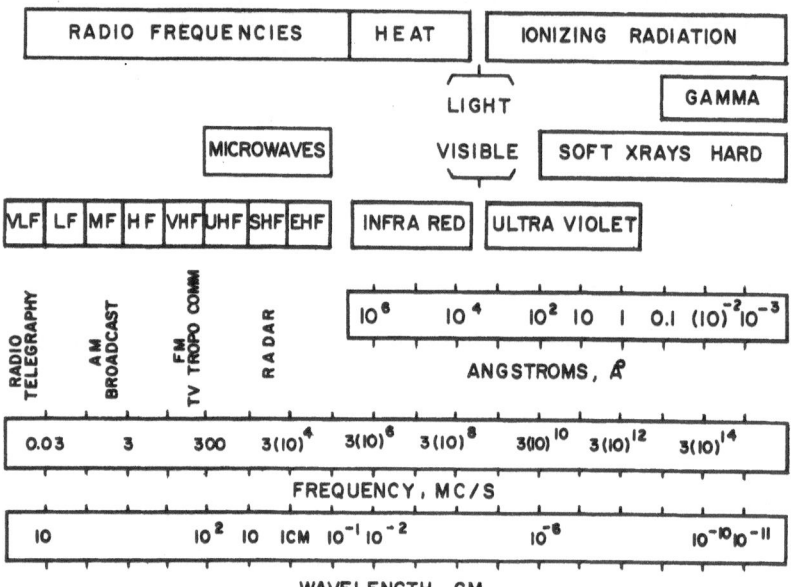

Fig. 1. Frequency spectrum.

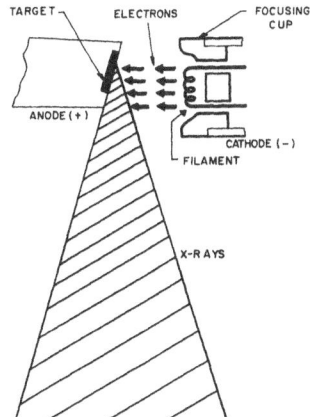

FIG. 2.   Schematic diagram: X-ray tube.

and thus forms the cathode, or negative electrode of the tube.   The sudden stopping of these rapidly moving electrons in the surface of the target results in the generation of X-rays.   Figure 2 is representative of the universal or conventional type X-ray tube used for medical and industrial purposes (1, 2).

High-power electronic tubes, such as klystrons, magnetrons, and high-voltage hydrogen thyratrons, possess the basic physical parameters which allows them to act as powerful X-radiation generators, i.e., a beam of electrons traveling at high speed towards an anode, or target, which is at a very high voltage and the subsequent stopping of these rapidly moving electrons.

Devices employing the aforementioned phenomena and operating with an applied voltage of more than 15 kv can be considered potential hazards.   In high-power microwave generators, conditions and components exist wherein the production of soft X-radiation at levels as low as 15 kv, through the "intermediate" range and on up to "hard" X-radiation at 300 kv is possible.   (Experimental efforts up to 400 kv have been common practice in the last few years.)

Note: Early workers in the X-ray field became aware of the fact that the continuous spectrum from X-ray tubes operating under higher applied voltages exhibited superior powers of penetrating materials.   Such penetrating X-rays were initially called "hard" X-rays.   Less penetrating radiations were classified as "soft."   A very loosely interpreted terminology known as the "intermediate" range was used for that portion of the spectrum between "soft" and "hard" X-rays.   These designations have been retained in use; however, it is now un-

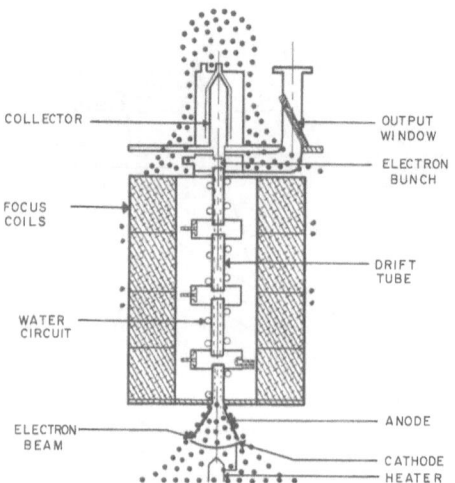

COLLECTOR

OUTPUT
WINDOW

ELECTRON
BUNCH

FOCUS
COILS

DRIFT
TUBE

WATER
CIRCUIT

ANODE

ELECTRON
BEAM

CATHODE

HEATER

FIG. 3. X-ray output—klystron amplifier.

derstood that "hard" radiation is more penetrating because it is shorter in wavelength (or higher in frequency or in quantum energy).

Figures 3, 4, 5, and 6 illustrate a pictorial concept of the theoretical and actual output of such X-ray producers as klystrons, traveling wave tubes, magnetrons, and thyratrons. These figures are a general representation of the X-ray output and in themselves may or may not be peculiar to any one type of tube in a specific category. These data are based on actual measurements made at RADC laboratories

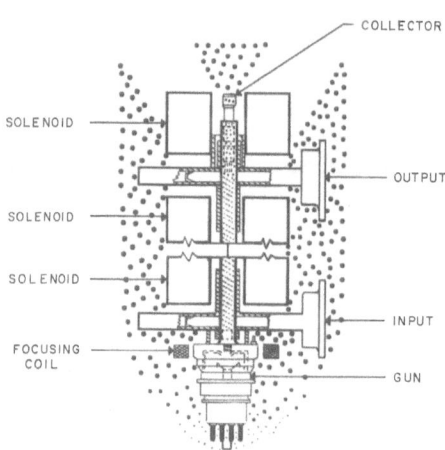

COLLECTOR

SOLENOID

OUTPUT

SOLENOID

SOLENOID

INPUT

FOCUSING
COIL

GUN

FIG. 4. Theoretical X-ray distribution—traveling wave tube.

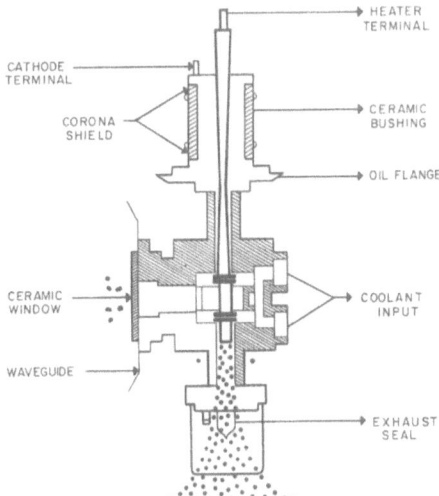

FIG. 5.   X-ray distribution—
          magnetron.

and commercial contractors' plants supplemented by a physical theoretical hypothesis relative to X-ray generation.   By the use of black dots, the artist has shown relative intensities and probable patterns which would depict the over-all X-ray output.   The thyratron and klystron represent the extremes in energy levels as well as intensity levels and shall be discussed in some detail here.

The hydrogen thyratron consists of an indirectly heated cylindrical cathode with an inner and outer cathode shield and cathode baffle.   The nickel grid structure consists of a cylindrical portion and a perforated grid disk with a solid grid baffle below it.   The molybdenum anode is enclosed by the grid disk.   The tube ordinarily is operated in a line-type pulse-modulator circuit with resonant charging of the pulse-forming network to a high peak forward voltage (3).

Let us assign the operating parameters of a hydrogen thyratron tube 1257 to our discussion.   These would be; 38 kv peak voltage, 2000 amp peak current with a 2.5 μsec current pulse at a repetition rate of 200 pulses/sec.   In order to get an idea of average currents for the production of X-radiation the duty cycle will be used, resulting in approximately 1 amp average current.   While the effective energy of this tube is rather low, 25 kev, the 1 amp current would give rise to rather intense "soft" X-radiation.   Again returning to the artist's conception of the thyratron, Figure 6, it is noted that the

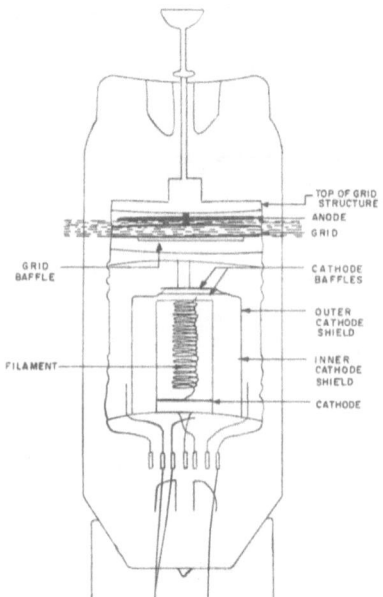

TOP OF GRID STRUCTURE
ANODE
GRID
GRID BAFFLE
CATHODE BAFFLES
OUTER CATHODE SHIELD
FILAMENT
INNER CATHODE SHIELD
CATHODE

Fig. 6. X-ray distribution—hydrogen thyratron.

major portion of the X-ray beam emanates in a circle through the screen mesh of the grid-anode region.   There may be considerable latitude in X-ray radiation from tube to tube under similar operating conditions due to variations in grid emission.   It has been reported that the 1257 has emitted X-radiation intensities up to 10,000 mr/hr at a distance of 1 ft from the tube.   On an average however, 1200–1500 mr/hr would be a more common radiation intensity.

For the most aggravated conditions 1/16 in. steel paneling would attenuate this radiation down to a fraction of 1 mr/hr.   Leaded glass 1/4 in. in thickness would attenuate 10,000 mr/hr to a negligible quantity.

Briefly, the high-power klystron operates on the principle of velocity modulation.   (It is this velocity modulation that is required for klystron bunching action.)   It consists of an electron gun, an rf section made up of a series of resonant cavities with drift tubes interposed, followed by an electron collector.   The electron beam generated by the electron gun is focused through the rf section, usually by means of a magnetic field.   The electrons are then dissipated in the electron collector.   These phenomena are represented in Figure 7.

Essentially these basic physical conditions existing in an ordinary X-ray tube — a beam of electrons traveling a high speed towards an anode, or target (in this case the collector), which is at a very high voltage and the stopping of these rapidly moving electrons — are present in the klystron.

In a klystron two conditions of X-radiation exist — one without rf applied to the tube and the other with rf applied. With the high voltage impressed on the tube and without the rf drive the beam of electrons is more concentrated throughout the body of the tube and less likely to impinge on many random targets giving rise to a multiplicity of X-radiation beams. X-radiation intensity is always found to be greater and more in evidence when the rf drive is impressed because of two probable reasons. The rf voltage at the output gap adds to the dc beam voltage, giving rise to more intense X-radiation, and the interposition of the rf field spreads the beam of electrons apart as it travels down the body of the tube.

These phenomena manifest themselves more readily when shielding considerations are required in order to make a klystron safe for personnel in its immediate proximity. For example, an experimental klystron has been designed which will give a peak output of 20 Mw. The peak voltage applied was approximately 400 kv with peak current approximately 250 amp. The average current evolved was approximately 30 ma (4).

From available data one would predict that a klystron operating at 400 kv and 30 ma average current will produce about 1100 r/hr. To reduce this level to the desired maximum permissible level as allowed by USAF regulations, 2.5 mr/hr for a man working in the vicinity of the tube 40 hr a week, would require an attenuation factor of about 4.5×10^5. Using information in available tables, it is found that better than 3 cm of lead is required (5).

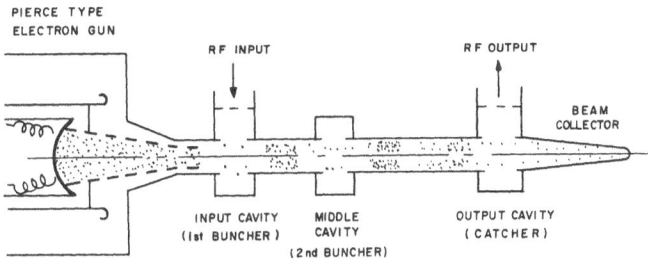

Fig. 7. Schematic representation of a typical cavity klystron.

This analysis is much too simple, and because we are concerned with a special type of X-ray generator, the klystron, the above conclusions must be modified. Conventional tables predict X-ray intensity on the basis of monochromatic radiation. The klystron produces a continuous X-ray spectrum and most of the radiation is less penetrating than those that correspond to the limiting 400 kv energy. The second consideration has to do with the fact that an oscillating klystron has groups of electrons accelerated to voltages approximately twice the beam voltage, or in this case 800 kv. X-radiation of greater intensity and greater penetrating powers would therefore be produced. The aforementioned phenomena have made the problem somewhat complex, as the velocity and space distribution of electrons in an oscillating klystron cannot be predicted with any degree of accuracy in the region beyond the third cavity. It follows that at best the calculated thickness of lead can only be used as a guide and the determination of safe radiation protection be accomplished by actual experimentation. Figure 8 therefore represents the required shielding on the tube described.

It is interesting to note that in the early stages of tube development and shielding for the AN/FPS-7 it was necessary to purchase

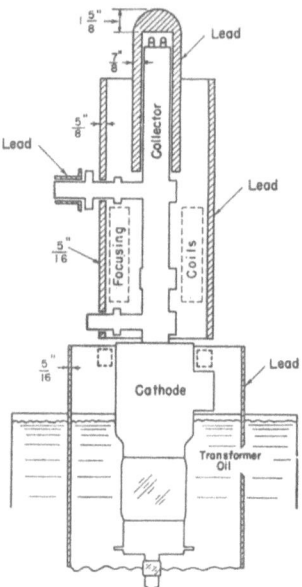

Fig. 8. Proposed shielding of klystron indicating location and thickness of lead.

a lead-shielded room. Lead thickness sufficient to attenuate 250 kv energies was supplied. Upon operating an unshielded tube within the lead room it was found the room literally leaked like a sieve. Closer examination revealed that levels as energetic as 480 kev were being produced. A special lead cup placed over the collector, the "hottest" spot on the klystron, allowed further use of the room.

In a tunable klystron, X-radiation output may be increased or decreased by going through the frequency range. Physical location of the X-radiation can be changed by beam focusing and "peaking" with variable parameters.

Detection

The detection of the X-radiation produced by rf tubes is a somewhat more difficult problem than the detection of similar radiation from standard X-ray equipment or from radioactive substances. One cause of this difficulty is that the X-radiation is produced simultaneously with an intense field of rf radiation. The rf radiation may produce strong potential gradients within the X-ray detector and its associated circuits, resulting in an erroneous indication of the X-ray intensity. The detector and circuits can be shielded against the rf radiation, but unless the detector has an inherently low sensitivity to rf radiation, the shielding required may greatly attenuate the lower energy X-radiation before it reaches the detector, making the detecting instrument insensitive to the lower energy X-rays. This is particularly important in the detection of X-radiation from rf tubes, where X-radiation having energies as low as 20 kev must be detected. Most conventional detectors are already limited by lower energy response by virtue of their basic design and almost none will respond below 50 kev (6).

Another source of difficulty in detecting X-radiation from rf tubes is that the X-radiation is generated in very short pulses of extremely high intensity, while the average X-ray intensity may be relatively low. For example, a tube such as a klystron may be operated at a duty cycle of 0.001, and produce an average X-ray dose rate of 1 r/hr a few feet from the tube. The intensity produced in a single X-ray pulse, however would be 1000 r/hr at the same distance from the tube. High instantaneous intensities such as these make conventional gas-filled X-ray detectors such as Geiger tubes and ionization chambers difficult if not impossible to use as reliable detectors of

X-radiation from rf tubes. This is so because intense radiation generates enough ion pairs in the gas of these detectors to seriously alter the electric field within the detectors, changing their sensitivity and possibly rendering them inoperative.

Additional considerations which must be taken into account evolve from other limitations of the previously mentioned instrumentation: energy response, size of radiated beam, and resolution.

Most X-ray emitting electronic tubes generally produce beams that are small, well collimated (particularly when shielded) beams through small faults or openings in the tube body or shielding. An intense beam 1/2 inch in diameter would not totally fill the volume of a 3-in. diameter ionization chamber. By definition the roentgen shall be that quantity or dose of X-radiation for which the associated corpuscular emission per 0.001293 g of air produces ions carrying 1 esu of quantity of electricity of either sign in air. The scale graduations of the ionization chamber have been calibrated against this definition, which requires a uniform X-ray field completely filling the volume of the chamber. A small beam would ionize only a portion of the volume of the chamber and this ionization action would be averaged throughout the entire volume of the instrument. Under these conditions an erroneous reading would result.

Another limitation of the ionization-chamber-type instrument to be considered is the attaining of true readings at the lower limit of pulse widths, and also at the upper limit of pulse widths. An ionization chamber could be adjusted so that the RC of the instrument would operate at narrow pulse widths, but as it approaches the wider pulse widths, operation would be doubtful. It would be difficult to determine whether the instrument was operating as an ionization chamber or an electron chamber. Add to this consideration another variable, the varying repetition rate from one electronic device to another, and it would seem that conditions become almost impossible (6).

The Geiger-Muller type device in addition to previously mentioned limitations has a relatively low resolution. Its recovery time is in the 10–100 μsec range and over, and therefore, even under the best conditions, the output signal for pulses over 10 μsec would not be proportional to the pulse width and no calibration of the instrument would be possible. The G-M tube has some use when measuring low intensity radiations, but in a very intense field, because of the low resolution, gives systematically low counting rates (7).

Photographic dosimetry, actually in use since Wilhelm K. Röntgen accidentally registered an invisible radiation on a photographic plate, still enjoys widespread use as a form of X-radiation detection. Photographic films, particularly where enclosed in a suitable badge, can also be used for personnel monitoring. Larger films placed in the environment of high-power generators not only can indicate levels of radiation intensity but describe patterns as well. Well-trained personnel can readily make energy level determinations given the time and proper ancillary equipment.

As a personnel monitoring device, photographic techniques do have several limitations: (a) Readings are not readily available; (b) developing and analysis of exposure must be accomplished under controlled conditions; (c) wide differences may occur in the number of roentgens required to produce a given density; (d) when used as a film badge, the acceptable area is so small that the probability of a narrow intense beam striking the badge is very remote.

Thus far the discussion has been of instruments which all have apparent limitations. One which has not been discussed is the scintillation-type detector, a comparative newcomer to the field for use as a radiation survey instrument. Basically the theory of operation is briefly as follows: X-radiation impinges on a crystal, the crystal fluoresces producing light in the visible spectrum, this light emission is in turn detected by a photo cell whose electrical output serves as an input to electronic circuitry where the signal is amplified and then fed into some device for visual indication. This is illustrated in Figure 9.

It became apparent that the use of a scintillation-type detector in

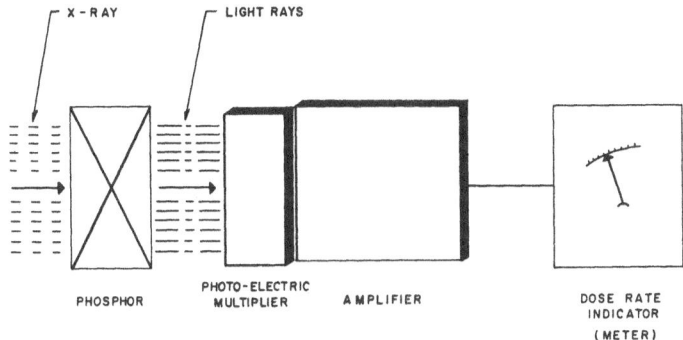

X-RAY   LIGHT RAYS

PHOSPHOR   PHOTO-ELECTRIC MULTIPLIER   AMPLIFIER   DOSE RATE INDICATOR (METER)

FIG. 9. Schematic diagram of scintillation-type detector.

conjunction with an integrating circuit might readily lead to the solution of the Air Force's problem. Certainly this was a technique where the detector was unaffected by rf interference, dynamic intensity levels, varying duty cycles, and the size of the detected beam. High-gain electronic circuits could be kept to a minimum by use of a photo-multiplier tube optically coupled to the pickup crystal. Actually, although not necessary, such a device could be built which could resist several w/cm² of rf energy.

The Radiacmeter ME-118, a scintillation-type device which will adequately detect and measure instantaneously a dose rate of pulsed X-radiation, has been developed by Rome Air Development Center. This instrument, with adjustment to its lower energy response, will be made available to the field and become a standard item in the Air Force inventory (8).

Summary

We have discussed thus far the physical concepts of generation and detection of pulsed X-ray from microwave sources. Now let us dwell momentarily on how all this is related to man. It follows that we must have some standards, some rules, a procedure or procedures to follow. Radiation protection standards do not readily permit direct testing, hence are not standards at all. They are merely over-all radiation protection guides!

It can readily be seen that most of our radiation protective criteria are based on absence of the demonstration of any deleterious influence. Because of our limited knowledge of the effect of low levels of radiation on animals and the almost non-existent knowledge of the effects of small doses of radiation on man, the permissible exposure levels for radiation values have been set low enough so that there is a negligible probability of radiation damage occurring (9).

Figure 10 illustrates how the recommended maximum permissible dose has dropped through the years. The early figures are mere approximations because it must be borne in mind that up to 1928 we had no acceptable unit of radiation dose. By the early 1930's the roentgen had become established as an acceptable unit, in terms of which the permissible dose level could be expressed. The value turned out to be approximately 0.2 r/day.

We must agree that man in his quest for higher and higher output power from his microwave generators has almost outdone him-

SOURCE	r / day	r / week	rems / yr
1902 ROLLINS	10		
1925 MUTSCHELLER	0.2		
1925 LEWIS	0.2		
1927 ADVISORY BOARD (HOLLAND)			
1928 BARCLAY & COX	0.17		
1931 ADVISORY COMM.(U.S.)	0.2		
1936 ADVISORY COMM.(U.S.)	0.1		
1950 ICRP		0.3	
1957 NCRP			5

Fig. 10. Recommended maximum permissible doses. [Taken from Stone, R. S., *Radiology, 58,* 639, 1952]

self. X-radiation from high-power microwave generators is in all respects a potential hazard. From all present indications it would appear that the present offenders, klystrons, magnetrons, etc., are not headed for extinction. It is very possible that the energy peak has been reached but the current consideration, the intensity factor of X-radiation, is progressively climbing. Where one tube alone will not give sufficient rf output, two, four, and even eight are being used in combination to reach unheard-of power levels. X-radiation levels of course will add up with the multiplicity of tubes.

Let us look with some optimism on the situation instead of total pessimism. Despite potential X-radiation outputs of unheard-of proportions, the USAF has equipments operating in the field which are relatively personnel safe. X-radiation is on the average so low that in some instances it is difficult to detect. Safety features are being designed into equipment which will prevent personnel from experiencing any catastrophic amounts of X-radiation. It must be noted, however, that all effects due to human frailty and ingenuity cannot be totally designed out of equipment.

As long as we have a thorough understanding as to the nature of this new potential hazard, detect and measure it with a fair degree of accuracy, and are willing to adequately shield our equipments, personnel should not be subject to harmful radiation. Interlock systems can be improved which would make the removing of protective shielding and subsequent operation of the equipment almost impossible.

Since the human system, as such, is not capable of recognizing the reception of ionizing radiation, constant monitoring of equipment, people, and environment is necessary for the detection of potentially harmful ionization areas and the establishment of barriers. Each day we are making strides toward this accomplishment, such that no one of our personnel will knowingly or accidentally be subject to harmful radiation. While we are literally making our first gains in this area, we are in a position to implement an educational program wherein a basic understanding of prevailing conditions is accomplished, supplemented by the improvement of instrumentation and techniques.

References

1. Eastman Kodak, *Radiography in Modern Industry,* 2nd ed., Rochester, 1957.

2. Blatz, H., *Radiation Hygiene Handbook,* McGraw-Hill, Toronto, 1959.

3. Reich and Schneider, "X-ray Emission From High Voltage Hydrogen Thyratrons," *Proc. I.R.E., Vol. 43* (June 1955).

4. Chodorow, Ginzton, Neilsen, and Saukin, "Design and Performance of a High Power Pulsed Klystron," *Proc. I.R.E., Vol. 41* (November 1953).

5. Astin, A. V., *X-ray Protection Handbook,* National Bureau of Standards, Washington D.C., 1955.

6. Price, W. J., *Nuclear Radiation Detection,* McGraw-Hill, Toronto, 1958.

7. Barnes and Taylor, *Radiation Hazards and Protection,* George Newnes Ltd., London, 1958.

8. Final Report Radiacmeter ME-118/G AF30(602)1692. RADC-TR-59-74. ASTIA-215 859.

9. Taylor, L. S., "Radiation Protection Standards," *Radiology 74,* 824, 1960.

Quick Formulas for Radar Safe Distances

H. S. OVERMAN
Guided Missile Safety Staff
U. S. Naval Weapons Laboratory
Dahlgren, Virginia

RADARS USED for weapons control aboard Naval missile ships are narrow-beam equipments using paraboloidal or microwave lens antennas with roughly circular apertures. The peak rf power radiated by these antennas may range up to several megawatts, and average power may range to several kilowatts. The antennas are located near the weather decks and are relatively unrestricted in rotation. There is thus a real possibility that personnel may be exposed in the beams. With some radars the power density on the beam axis exceeds the nominal "safe" value of 10 mw/cm² at distances of several hundred feet, which may be well beyond the physical structure of the ship.

In assessing the safety of specific radar installations we frequently need to estimate the maximum distance to which the 10 mw/cm² level will extend, under essentially free-space conditions. The method for making these estimates preferably should be fast to use, should involve a minimum of mathematics, and should yield results in reasonably close agreement with antenna theory. In this paper is presented a set of formulas which meets these requirements. From these formulas it is possible to compute directly the "safe" distance from three quantities which are readily determined: (1) the diameter, in feet, of the antenna aperture; (2) the average rf power, in watts; (3) the wavelength, in centimeters.

These formulas are listed in Figure 1. Three different sets of computations are indicated in this figure. Only one set, however, need be applied in any one specific case. The proper one to apply is

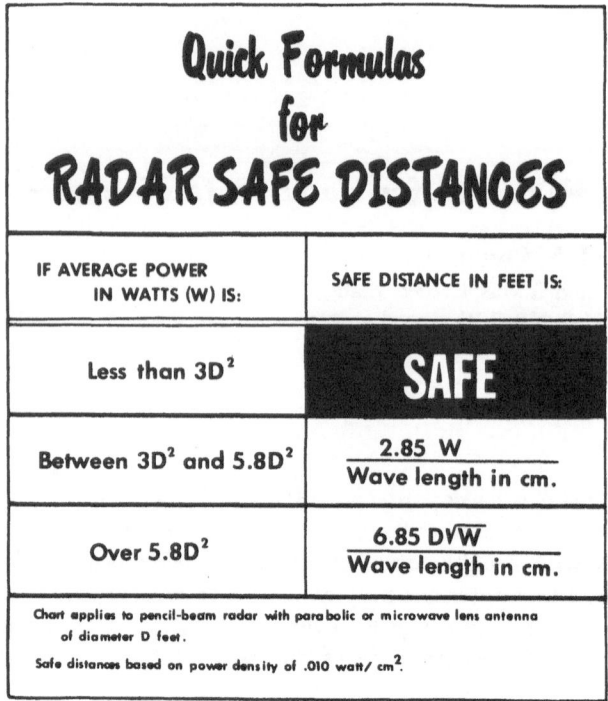

FIG. 1.   Safe distance formulas.

dependent on the ratio of average power to the square of the antenna diameter, and is determined from the first column of the figure.

As an aid to understanding the use and limitations of the formulas, it is of interest to consider some characteristics of radar beams. Figure 2 is a much simplified representation of a radar beam which is useful for visualizing the conditions at a considerable distance away from the antenna, in what is generally called the "far field." The power flow in such a beam may be thought of as confined within a cone which has its apex at the antenna. The apex angle of the cone is the "beam width." The cross-sectional area of the beam varies with the square of the distance from the antenna; hence the power density, which is the power per unit area, will be proportional to the reciprocal of the square of the distance, i.e., it will conform to the inverse square law. (In practice, the power radiated by the antenna is not all confined within the conical beam. Some is radiated just outside the nominal limits of the beam, and some in side lobes.

Moreover, the power density is about twice as great on the beam axis as at the edges. These deviations from the "simplified" case need not concern us here, except to note that they must be taken into account in any quantitative computation made on the basis of radiated power and antenna geometry.)

Now, the formula in the bottom row of Figure 1 is simply a re-statement of the inverse square law. It states that for a given antenna size and wavelength, the distance to the 10 mw/cm² power density level will be proportional to the square root of the average radiated power. If the beam were conical all the way to the antenna, as in the simplified case described, this would be the only formula needed. Actually, however, the conical beam does not exist at the antenna. The antenna is designed so that, geometrically, energy will be transmitted in parallel rays. The power leaving the antenna is thus confined within a cylinder having the same diameter as the antenna, as shown in Figure 3. (We will assume, for the discussion which follows, that the antenna is uniformly illuminated by the feed.) Since the cross-sectional area of this "near-field" beam is independent of distance, it would seem reasonable to expect that the power density throughout would be the quotient of the transmitted power by the

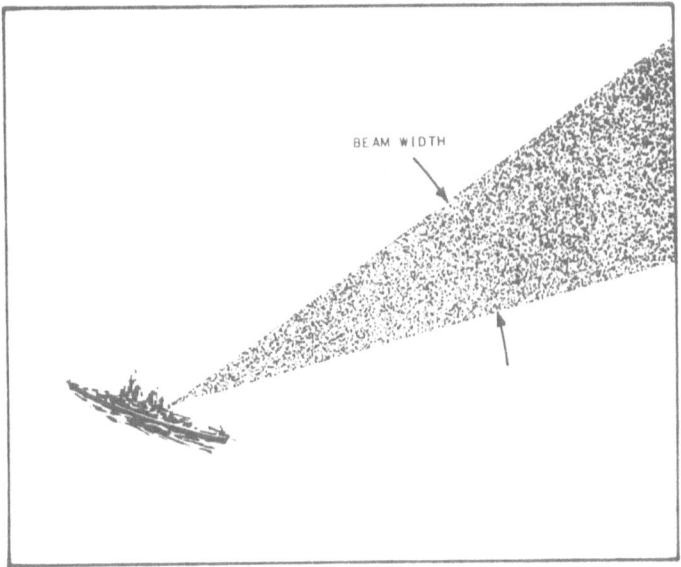

FIG. 2. Radar beam, simplified representation.

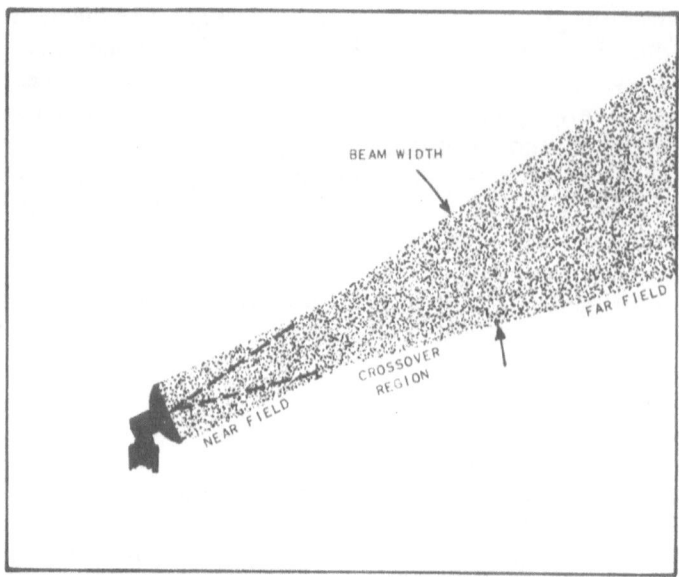

Fig. 3. Radar beam, showing conditions near antenna.

antenna area. This is not true, however. Because of an optical
phenomenon, "Fresnel diffraction," there are many peaks and valleys
in the power density contour in the near field. Figure 4 shows
a model of the complex power density distribution across the beam
at a number of distances along the beam axis. Fortunately, it is not
necessary to calculate, point by point, the values of these peaks and
valleys in order to draw conclusions about safe distances. From the
safety standpoint, we are more interested in the maximum peaks
within the near field. These always occur on the beam axis and have
a value which is theoretically four times the quotient of the power by
the antenna area. The last of these peaks occurs at a distance equal
to the square of the antenna radius divided by the wavelength,
a point which, for our purposes, may be defined as the limit of the
"near" field. It thus can be seen that if the power density at the end
of the near field is less than the predetermined "safe" value, the entire beam can be considered safe. This is reflected in the first row of
Figure 1.

Between the end of the near field and the beginning of the far
field, there lies a transition zone in which the power density decreases
with increasing distance, but not in accordance with the inverse
square law. This zone has been called the "crossover region." In the

crossover region a fair approximation to the safe distance can be made on the assumption that the safe distance is directly proportional to the average radiated power, resulting in the formula shown in the middle row of Figure 1.

The relationship between the simplified formulas and antenna theory is as follows:

The power density as a function of distance on the axis of a radar beam under uniform antenna illumination may be computed from the following equation:

$$p = (4WK/\pi a^2) \, [\sin (\pi a^2/2\lambda r)]^2 \qquad (1)$$

where p = power density
r = distance from antenna
a = radius of aperture
λ = wavelength of radiation
W = average power delivered to antenna
K = an empirical correction factor lying between 0 and 1

If it is assumed that there are no losses, K takes on the value 1.0 and eq. (1) is then identical with the theoretical result derived by Silver (1) for the uniformly illuminated microwave antenna. In practice, K will seldom exceed 0.60. It is assumed to be 0.60 throughout the remainder of this discussion.

For small values of the distance r in eq. (1), the sine-squared term oscillates between 0 and 1.0, resulting in a series of peaks and valleys spaced progressively closer as the antenna is approached. This is the region we have previously called the near field. The maximum

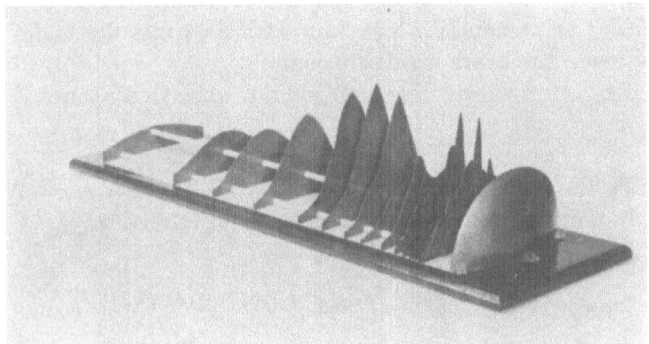

FIG. 4. Model showing distribution of power density in a plane containing the beam-axis of a microwave antenna.

power density in this region is given by the height of these peaks, which is, from eq. (1),

$$p_{(near)} = 4WK/\pi a^2 \tag{2}$$

For quite large values of the distance r, the sine of the quantity $\pi a^2/2\lambda r$ can be considered equal to the quantity itself to a first approximation. Under this condition, eq. (1) reduces to the inverse square law of the far field:

$$p_{(far)} = WK(\pi a^2/\lambda^2 r^2) \tag{3}$$

In the intermediate distance between the near field and the beginning of the far field, both eqs. (2) and (3) yield values of p which are too high. The following empirical equation provides a much better approximation to the theoretical result in the intermediate region:

$$p_{(intermediate)} = 0.87(W/\lambda r) \tag{4}$$

Figure 5 is a logarithmic plot of eqs. (2), (3), and (4) together with the theoretical result (eq. (1)) for comparison. In Figure 5, distances from the antenna are expressed in terms of the dimensionless quantity $r\lambda/D^2$. At the intersection of the lines representing eqs. (2) and (4), this quantity has the value of 0.284. Equations (3) and (4) intersect at 0.541. These intersections define the limits within which each of the three equations yields a valid approximation. They are roughly coincident with the limits of the near, intermediate, and far fields.

Referring again to Figure 5, in the region where eq. (2) is valid the quantity pA/W, where A is the aperture area, has a constant value of 2.4, from which the maximum value of p is seen to be $3.06W/D^2$; i.e., when W is less than $0.327D^2$ times the "safe" power density level, the beam is safe throughout.

In the region in which eq. (4) is valid, the safe distance

$$r_{(safe)} = 0.87W/\lambda p_{(safe)} \tag{5}$$

and since the dimensionless distance $r\lambda/D^2$ lies between 0.284 and 0.541, W will lie between $0.327D^2 p_{(safe)}$ and $0.622D^2 p_{(safe)}$.

In the region in which eq. (3) is valid

$$r_{(safe)} = (a/\lambda)\sqrt{\pi WK/p_{(safe)}} = 0.685(D/\lambda)\sqrt{W/p_{(safe)}} \tag{6}$$

and W is, of course, equal to or greater than $0.622D^2 p_{(safe)}$.

The foregoing analysis has been based on a consistent system of units. Specifically, if power density is expressed in watts per square

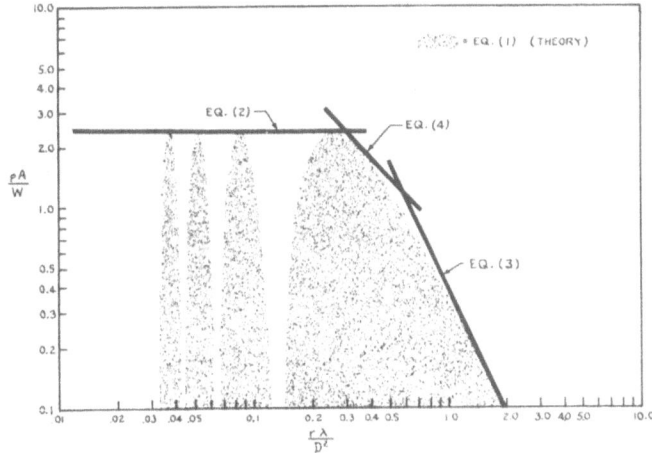

FIG. 5. Power density vs. distance on the beam-axis; theoretical and approximate curves.

centimeter, then in order to obtain correct numerical results, λ, D, and r must be expressed in centimeters. While it is usual to express power density and wavelength in watts per square centimeter and centimeters, respectively, the antenna dimensions and distance are usually expressed in feet. Conversion of the foregoing formulas to the latter terms, and assigning a numerical value of 0.010 w/cm² to the safe power density, $p_{(safe)}$, result in the following relationships:

(a) If the average transmitted power in watts, W, is less than $3.03D^2$, where D is the antenna aperture diameter in feet, the beam is safe, i.e., the power density does not exceed 0.010 w/cm² at any point.

(b) If W lies between $3.03D^2$ and $5.78D^2$, the safe distance in feet $r_{(safe)}$ is

$$2.85W/\lambda \qquad (7)$$

(c) If W exceeds $5.78D^2$, the safe distance in feet is

$$6.85(D/\lambda)\sqrt{W} \qquad (8)$$

No great loss in accuracy will result if the values of W/D^2 are rounded off to 3.0 and 5.8 in lieu of 3.03 and 5.78, respectively. This has been done in the safe distance chart, Figure 1.

There are certain limitations which should be kept in mind when using these formulas to evaluate the personnel hazard in a practical application. The first, and most important, lies in the fact that the safe distances predicted by the formulas apply only to equipments operating at optimum practical efficiency. Poor transmitter tuning or transmission losses higher than the design value, for example, tend to reduce the distance to which the beam is hazardous. It has been a common experience at the Naval Weapons Laboratory to find the measured power density some 3 db down from the predicted value.

The second limitation lies in the alternative possibility that the power density at some point in the beam could exceed the predicted value, either because of a particularly efficient design in which the factor K exceeds the assumed value of 0.60, or because of in-phase reflections from objects near the beam. Neither of these conditions is considered to be of particular concern. The first can result, at most, in a 25% increase in power density, and the second is so limited in space coordinates and transient in duration as to present only a remote probability of exposure to levels significantly higher than nominal.

A third, and generally negligible limitation arises from the fact that the formulas are derived from a theory based on uniform illumination of the antenna aperture, whereas in practice the illumination is usually tapered, being less at the edge of the dish than at the center. Mumford and Engelbrecht (2) have shown, however, that tapers up to 10 db applied either as a linear or square law function of the dish radius will have a negligible effect on either the on-axis power density in the far field or on the maxima in the near field. In the case of extremely high tapers, of the order of 25 to 35 db, Hansen and Bailin (3) have shown that near-field maxima closer to the antenna than the first maximum may be somewhat higher than values computed on the assumption of uniform illumination.

References

1. Silver, S., *Microwave Antenna Theory and Design,* McGraw-Hill, New York, 1949, p. 199.

2. Mumford, W. W., and Engelbrecht, R. S., "Microwave Radiation Hazards: Power Density Along the Axis of a Circular Antenna with Tapered Illumination," *Bell Telephone Laboratories Technical Memorandum No. MM-58-6635-7,* Whippany, New Jersey, Oct. 6, 1958.

3. Hansen, R. C., and Bailin, L. L., "Near Field Analysis of Circular Aperture Antennas," *Hughes Aircraft Company Report No. AFCRC-TN-59-780,* Culver City, California, Aug. 1959.

Some Engineering Aspects of Microwave Radiation Hazards*

R. S. Engelbrecht and W. W. Mumford
Bell Telephone Laboratories
Whippany, New Jersey

GENERAL

WE ARE currently quite acutely aware of the fact that a potential health hazard exists in the vicinity of some of the high-powered radars in use today and that the future will bring even higher powers. In order to forestall the possibility of casualties in the future, we are establishing exposure limits and are designing our equipments accordingly.

It is logical that our earlier efforts were directed toward the establishment of reasonable limits. Although much has been published on this subject and tentative limits have been established, we still need more information on many facets of the biological effects of the microwave radiation. Nonthermal effects which may be associated with high peak pulse power are suspected but little is known as yet. Testicular change and chromosomal aberration have been reported in the literature, but corroborative and more extensive tests are needed. The duration and wavelength of the exposure may eventually be built into our exposure limits when more data are acquired. These are but a few examples to illustrate the point that we do need more information on biological effects. This is being supplied regularly at the Tri-Service meetings.

In the meantime, the radar design engineer accepts the currently adopted limits and attempts to design the new equipments to meet those specifications. Before the design of the new equipment is on

* An excerpt from a more complete discussion to be published elsewhere, possibly in the *Proceedings of the Institute of Radio Engineers*.

the drawing boards, the designer must be able to estimate the amount of radiation from the proposed apparatus and to incorporate into his design the appropriate protective devices and shielding so that the completed installation shall meet the requirements dictated by the exposure limits, whatever these limits may be. (The currently adopted exposure limits are based upon the average power density.)

This paper presents some simple approximate formulas for calculating the average power density along the beam of conventional antennas and points out their limitations. It also presents a new and more reliable formula for estimating the shielding effect of a metallic screen. A nomograph based upon this formula is submitted.

Calculation of Power Densities

Transmission formulas are required for calculating the power density in the beam of an antenna. These may be found in the literature (1, 2), but simpler approximations are needed.

The field in front of the usual parabolic antenna can be characterized by referring to two separate regions:

(1) The "near-field" or Fresnel region, where the radiation is substantially confined within a cylindrical pattern.

(2) The "far-field" or Fraunhofer region, beyond the Fresnel region in free space, where the radiation is essentially confined to a conical pattern and the power density along the beam axis falls off inversely with the square of the distance.

To compute the value of the power density in the near field, assuming a circular "dish" antenna, use:

$$W = 16P/\pi D^2 = 4P/A \tag{1}$$

where P = average power output, *not* peak power
        D = diameter of antenna
        A = area of antenna

If this computation reveals a power density which is less than the limit, then there is no need to proceed with further calculations, since eq. (1) gives the *maximum* power density that can exist on the axis of the beam of a properly focussed antenna. (A de-focussed antenna could give more, but that condition is not usual.)

If the computation from eq. (1) reveals a power density greater than the limit, then one assumes that this value may exist any place in the near-field region, and attention is directed to the far-field region.

In the far-field region, the free-space power density on the beam axis may be computed from

$$W = GP/4\pi r^2 = AP/\lambda^2 r^2 \tag{2}$$

where λ = wavelength.

The distance from the antenna to the intersection of the near-field eq. (1) with the far-field eq. (2), is given by the expression:

$$r_1 = \pi D^2/8\lambda = A/2\lambda \tag{3}$$

These formulas do not include the effect of ground reflection which could cause a value of power density which is four times the free-space value.

Setting the power density equal to the potentially hazardous level of 10 mw/cm^2, one may calculate the distance to the boundary of the potentially hazardous zone. This has been done for some of the radars which have antenna apertures that are round. More involved calculations are necessary when the antenna shape and illumination are more complex. Other calculations have been reported for several types of radars (1) and these are included in Table I.

Let us see how the approximate eqs. (1), (2), and (3) were derived and just how accurate they are.

Assume that we have a plane electromagnetic wave impinging on a totally absorbing body. If the projected area of the body is A, then the power absorbed will be the power density times the area or

$$P = WA \tag{4}$$

If the field strength of the incident wave is E volts per meter in free space, then the power density in watts per square meter will be

$$W = E^2 \, \text{v/m}/377 \text{ ohms} \tag{5}$$

Ordinarily it is not necessary to specify the field strength in volts per meter, but only to specify the power density in watts per unit area.

For an isotropic radiator in free space, radiating a total average power, P, in all directions equally, the power density on the surface of a concentric sphere of radius r will be simply the total radiated power divided by the area of that sphere, or:

$$W = P/A = P/4\pi r^2 \tag{6}$$

TABLE I

Distance in Feet from Radar Antenna to Boundary of

Potentially Hazardous Zone*

(Arranged in descending order of distances)

Radar type	Distance, ft, for 0.01 w/cm²	Radar type	Distance, ft, for 0.01 w/cm²
AN/FPS-16		AN/MPQ-21	165
Sig C Mod.	1020	AN/TPS-1G	
Standard Mod.	590	(40′ x 11′)	150
AN/FPS-6	560	AN/FPS-36	150
Herc. Imp. Acq.		Ajax TTR	132
HIPAR (Fixed)	550**, ***	Herc. Imp. Acq.	
AN/MPS-23	530	(Fixed)	130**, ***
AN/MPS-14	472	Herc. Acq.	
Herc. Imp. TTR	400	(Fixed)	130**, ***
AN/TPQ-5	350	AN/FPS-14	109
AN/FPS-20	338	AN/FPS-4	
AN/MPQ-21 (10′)	300	(narrow pulse)	106
Herc. MTR (Ajax)	270	AN/MPS-8	
Ajax Acq. (Fixed)	260**	(narrow pulse)	106
AN/CPS-9	260	AN/TPS-10D	
AN/MPQ-21 (7′)	210	(narrow pulse)	106
AN/MPS-4	205	AN/MPS-10 (C)	105
AN/FPS-8		AN/FPS-8	101
(40′ x 14′)	205	AN/MPS-11	101
Ajax MTR	205	Scr 584	70
AN/CPS-6B	200	AN/MPQ-10 (S)	50
AN/FPS-10	200	AN/TPS-1-D	50
AN/MPS-22	185	AN/FPS-25	40
AN/FPS-18	178	AN/FPS-31	27.5
AN/MPS-12	175	Herc. Imp. Acq.	
AN/MPQ-18	175	HIPAR (Rot.)	25
AN/FPS-3	172	Ajax Acq. (Rot.)	8
AN/MPS-7	172	AN/PPS-4	2.5

* Based on the following assumptions: (1) Free-space transmission. (2) No ground reflections. These could double the distances shown. (3) Calculations apply to the axis of the beam, i.e., where the power density is greatest. (4) The beam is considered to be fixed in space, i.e., not scanning.

** Not normally used with fixed antenna.

*** Interlocks provide assurance that transmitter is idle unless antenna is rotating.

Now if the radiator is not isotropic and radiates with a directivity gain G in a given direction, then the power density at a distance r would be, in the far-field region:

$$W = GP/4\pi r^2 \tag{7}$$

Allowing for 100% ground reflection, which doubles the electric field strength (and hence quadruples the power density), we have:

$$W = GP/\pi r^2 \tag{8}$$

If we express the gain G in terms of the antenna area thus:

$$G = 4\pi A/\lambda^2 \tag{9}$$

and insert this in eqs. (7) and (8), we have the alternative expressions in terms of antenna area, (eq. (2)),

$$W = AP/\lambda^2 r^2 \tag{10}$$

for the free-space power density, and

$$W = 4AP/\lambda^2 r^2 \tag{11}$$

when allowing for 100% reflection from the ground.

It will be convenient to express the far-field free-space power density (eq. (10)) in terms of the power density at the antenna aperture, $W_0 = P/A$ and hence $P = W_0 A$. Substituting this expression for P in eq. (10) and dividing both sides by W_0, we have,

$$W/W_0 = (A/\lambda r)^2 \tag{12}$$

For convenience, this may be rewritten thus:

$$W/W_0 = 4(A/2\lambda r)^2 \tag{13}$$

This expression applies in the far-field region but a simple modification makes it applicable to the near-field region for a uniformly illuminated round aperture, thus:

$$W/W_0 = 4\sin^2(A/2\lambda r) \tag{14}$$

This more exact expression is plotted in Figure 1 along with the lines representing the approximate formulas, eqs. (1) and (2).

In this graph and in some subsequent graphs, the relative power density is plotted in decibels on the ordinate and the distance, or more specifically $\lambda r/A$, is plotted logarithmically on the abscissa.

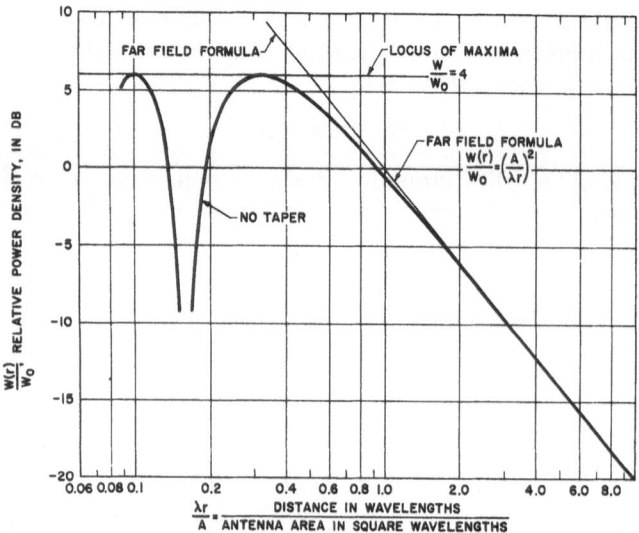

FIG. 1. Power density vs. distance for uniformly illuminated round aperture.

Note the alternate maxima and minima in the near field. The maxima all are 6 db above (4 times) the power density at the aperture. To be realistic in the near-field region we must assume that a man could be standing at such a distance that he could be exposed to a maximum. Hence the near-field approximation:

$$W = 4W_0 = 4P/A \tag{15}$$

This agrees with eq. (1).

Note that the near-field approximation intersects the far-field approximation when

$$r_1 = A/2\lambda = \pi D^2/8\lambda \tag{16}$$

as given in eq. (3).

Also notice that the largest deviation of the approximations from the more exact expression occurs at this distance and this deviation amounts to only 1.5 db which, in many cases, is of little consequence. For all other distances the error will be less than this 1.5 db for the conditions assumed above.

Thus it appears that approximations (1) and (2) are fairly reliable. They do, however, neglect the loss of power due to "spill over" at the antenna and the effectiveness of the antenna area and hence, in general, will give conservative estimates. Likewise, transmitter powers are often rated as power available *at the generator* so that any transmission line losses between the generator and the antenna would make the estimates even more conservative.

So far, we have assumed uniform illumination of a round aperture, whereas most antennas have the illumination tapered so as to reduce the side lobes. Often a square-law taper to 10 db down at the edge of the aperture is used. Let us see what this does to *the power density when the total radiated power is kept the same for different tapers.*

Figure 2 is a plot of the relative power density at the aperture of the antenna as a function of the radial distance from the center of the aperture for different tapers. For uniform illumination the relative power density is, of course, unity all over the aperture. However, for a linear electric field taper, the power density is more than three times this value at the center, tapering to about 0.3 at the edge. For a square-law taper of the electric field, the power density at the center of the antenna is over twice that for the uniform case and tapers to about 0.2 at the edge.

Judging from those curves, one *might* think that the power density maxima along the axis in the near field and the power density in the far field would be quite different for these three tapers. Such, however, is not the case as will be shown in the following analysis.

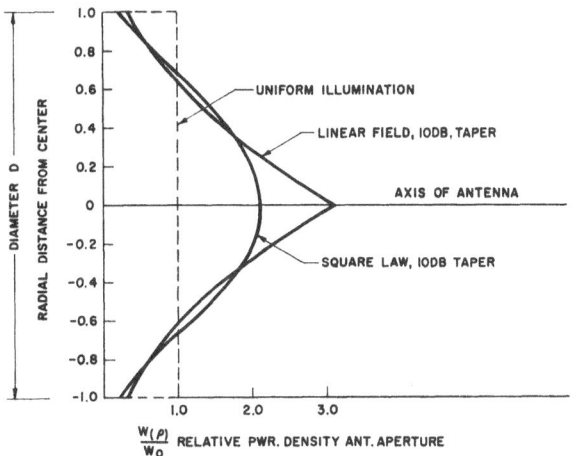

Fig. 2. Comparison of power densities at the aperture for different tapers of illumination.

We assume a circular aperture of diameter $2a$ with an illumination taper function $A(\rho)$. The field strength at a distance r along the axis of the antenna is given by the well known relation:

$$E_{(r)} = (i\beta E_0/r) \exp(-i\beta r) \int_0^a A(\rho) \exp(-i\beta(r' - r))\rho\, d\rho \quad (17)$$

where E_0 is the electric field at the center of the aperture,

$$\beta = 2\pi/\lambda$$

$$A(\rho) = \text{normalized amplitude taper}$$

If we now neglect the angular dependence of the Huygen's sources, then, for distances more than a few diameters away, the following approximation may be used:

$$(r' - r) = \rho^2/2r \quad (18)$$

$$(\text{for } a/r \ll 1)$$

Thus, when the distance is much greater than the radius, a good approximation is given by the relation:

$$E_{(r)} = (i\beta E_0/r) \exp(-i\beta r) \int_0^a A(\rho) \exp(-i\beta\rho^2/2r)\, \rho\, d\rho \quad (19)$$

For uniform illumination, $A(\rho) = 1$ and the formula quoted previously is derived.

For a linear taper

$$A(\rho) = 1 - (1 - \alpha)(\rho/a) \quad (20)$$

where
$$\alpha = E(\rho = a)/E(\rho = 0)$$

Performing the quadrature results in the following expression:

$$W(r)/W_0 = \frac{6}{1 + 2\alpha + 3\alpha^2}\left|1 - \alpha\cos(\pi/2)\mu_0^2\right.$$

$$\left. - \frac{1 - \alpha}{\mu_0}C(\mu_0) + i(\alpha\sin(\pi/2)\,\mu_0^2 + \frac{1 - \alpha}{\mu_0}S(\mu_0))\right|^2 \quad (21)$$

where
$$W_0 = P/\pi a^2$$

$$\mu_0^2 = 2a^2/\lambda r$$

C and S are the Fresnel integrals:

$$C(\mu) = \int_0^\mu \cos(\pi/2)\,v^2\, dv$$

$$S(\mu) = \int_0^\mu \sin(\pi/2)\,v^2\, dv$$

as tabulated in the book by Jahnke and Emde (6).

This expression is plotted in Figure 3 where it is seen that the major effect of the taper has been to fill in the deep nulls which existed in the near field for the uniformly illuminated case. The difference between this calculated curve and the approximate far-field eq. (2) is less than 1/2 db in the far field and again about 1.5 db at the intersection where $r = A/2\lambda$.

In the near field, the maxima are within about 1/2 db of the $W = 4W_0$ line.

For the case of a square-law taper of the electric field, the amplitude function becomes:

$$A(\rho) = 1 - (1 - \alpha)(\rho/a)^2 \tag{22}$$

Performing the indicated quadrature results in the following expression:

$$W(r)/W_0 = \frac{3}{1 + \alpha + \alpha^2}\Bigg|1 - \alpha\cos(\pi/2)\,\mu_0^2 - (1 - \alpha)\frac{\sin(\pi/2)\mu_0^2}{(\pi/2)\mu_0^2}$$
$$+ i\left(\alpha\sin(\pi/2)\mu_0^2 + (1 - \alpha)\frac{1 - \cos(\pi/2)\mu_0^2}{(\pi/2)\mu_0^2}\right)\Bigg|^2 \tag{23}$$

where again $\qquad\qquad W_0 = P/A$

and $\qquad\qquad\qquad \mu_0^2 = 2a^2/\lambda r$

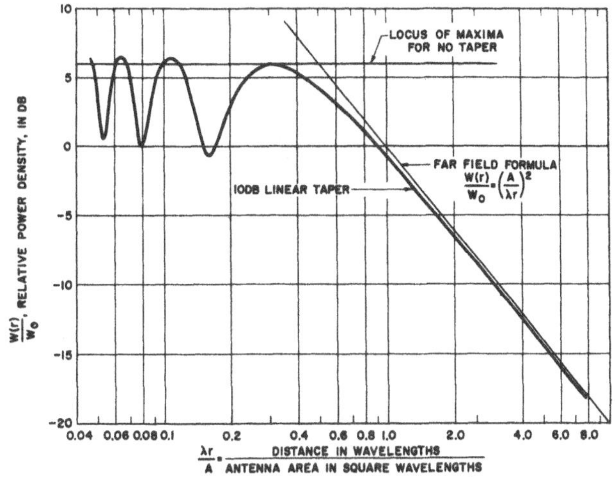

FIG. 3.   Power density vs. distance for round aperture with linear taper (10 db).

A plot of this expression is given in Figure 4. One curve and two straight lines are plotted. The straight lines are the approximations of eqs. (1) and (2). The curve is for a square-law taper of 10 db. It is seen that the departure from the approximate formula is not of much consequence.

Thus it is seen that tapering the illumination of a round aperture has not affected the power density on the axis enough to cause concern in either the far field or the near field.

What about other shapes of antennas, for example, a square antenna? The formula for a square antenna with uniform illumination is given by the relation:

$$W(r)/W_0 = 4[C^2(d/\sqrt{2\lambda r}) + S^2(d/\sqrt{2\lambda r})]^2 \tag{24}$$

where $W_0 = P/A = P/d^2$

This is plotted in Figure 5, together with the approximations based upon *equal area antennas*. For the square antenna the near-field maxima are lower than the approximation, but the one at the greatest distance is only about 1 db low. In the far field, the asymptotic agreement is excellent.

To summarize these observations, Figure 6 is presented. This

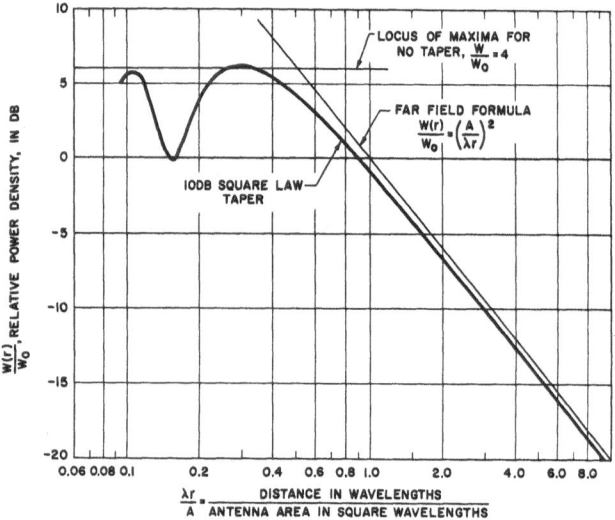

FIG. 4. Power density vs. distance for round aperture with square-law taper (10 db).

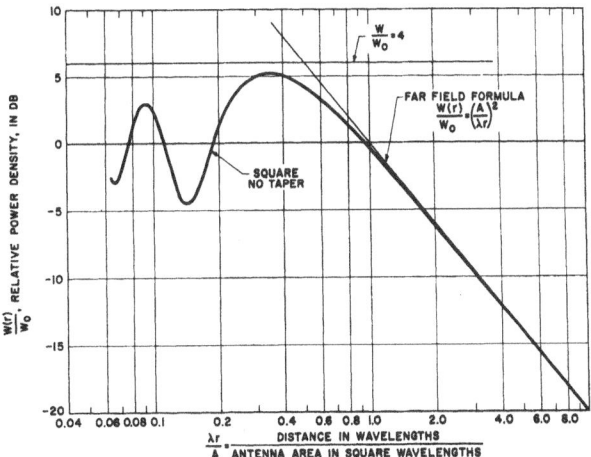

FIG. 5. Power density vs. distance for square aperture uniformly illuminated.

shows the near- and far-field approximations along with the maxima for the round antennas with linear and square-law 10 db tapers and for the square antenna with uniform illumination.

Thus the approximate formulas are seen to apply quite well to round and square antennas. For more exact results at the region of

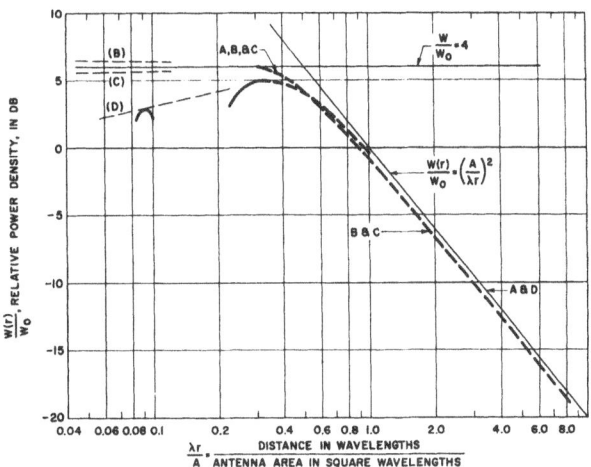

FIG. 6. Comparison of power densities for different shapes and tapers. (A) Round, no taper. (B) Round, 10 db linear taper. (C) Round, 10 db square-law taper. (D) Square, no taper.

crossover between near and far fields, the expression containing $\sin^2$ may be used (eq. 14).

For long rectangular antennas, an approximate formula is:

$$W/W_0 = (A/\lambda r)^2 \text{ beyond the distance } r > d_1^2/2\lambda \qquad (25)$$

and

$$W/W_0 = 2d_2^2/\lambda r \text{ within the distance } r \leq d_1^2/2\lambda \qquad (26)$$

$$d_1 = \text{wide dimension}$$

$$d_2 = \text{narrow dimension}$$

This near-field formula was derived from a more complete analysis for distances such that

$$d_1/r < 1$$

and

$$d_2^2/\lambda r < 1$$

Reviewing the foregoing discussion, it is seen that a simple calculation of the maximum power density in the near field according to eq. (1) reveals whether or not a hazard is involved in the beam of the antenna. If this power density exceeds 10 mw/cm^2, then a simple calculation based upon the far-field eq. (2) reveals the potentially hazardous distance in free space. These formulas apply to uniform illumination of square or round apertures or to illumination which is tapered in amplitude for round apertures. For other shapes and tapers a more complicated analysis would be necessary.

To calculate accurately the power density off the axis of the main beam requires the solution of a more difficult mathematical problem. One approach (1) reveals that the collimated beam in the near field falls off approximately 12 db per radius.

Many of our radars do not have the simple shapes nor simple illumination tapers that were treated above. In such cases the approximate formulas will not apply directly and a more complete analysis is indicated.

Scanning Antennas

The specified limits for safe and potentially hazardous power densities have been based upon average power. In the case of the scanning antenna, the average power absorbed by a fixed subject will be reduced by the ratio of the effective beam width to the scanned angle. Accordingly, the potentially hazardous distance is reduced by the square root of this ratio.

The effective beam width in the far field will, in general, be somewhat greater than the 3 db beam width and somewhat less than the width to the first null. The exact value depends, of course, upon the form factor of the radiation pattern.

In the near field the effective angle of the beam width will vary with distance since the field is collimated. Here the average power density of the scanning antenna is given approximately by the relation:

$$W \doteq W_0(D/2\pi r)\,(360/\theta) \tag{27}$$

for
$$\theta \geq (D/2\pi r)\,360$$

where θ is the scanned angle in degrees. For

$$\theta < (D/2\pi r)\,360°$$

$$W \doteq W_0 \tag{28}$$

Since
$$W_0 = 4P/\pi D^2 \text{ (circular aperture)} \tag{29}$$

$$W \doteq (4P/\pi D^2)\,(D/2\pi r)\,(360/\theta)$$

for
$$\theta \geq (D/2\pi r)\,360$$

and
$$W = 4P/\pi D^2 \tag{30}$$

for
$$\theta < (D/2\pi r)\,360$$

Setting $W = 10 \text{ mw/cm}^2$, the potentially hazardous distance in the near field is:

$$r_{\text{(hazardous)}} = (1/5\pi^2)\,(P/D)\,(360/\theta) \text{ cm} \tag{31}$$

for
$$\theta \geq (D/2\pi r)\,360 \quad \text{and} \quad r \leq D^2/4\lambda$$

where P is in milliwatts, D is in centimeters.

For
$$\theta < (D/2\pi r)\,360$$

the previous discussion is applicable.

Having once established the boundary of the potentially hazardous zone, where power densities of 10 mw/cm² may occur, appropriate measures must be adopted to keep personnel from entering that zone. Barricades or shielding fences may be installed. If gates must be provided for occasional access to the area by personnel, interlocks should be provided to ensure that the transmitter be idle whenever personnel open the gate and enter the area.

In some cases it may be necessary that the transmitter be kept operative while personnel enter into an area where the field is potentially hazardous. In this event, some kind of shielding arrangement is indicated. Whether this be of the fence or shielded room type or the mobile-portable shielding garment type, some knowledge of the effectiveness of the shield is necessary in order to be sure that the power density within the shielded area is low enough to be safe. Of course, completely enclosing the area in a water-tight copper container would most surely be adequate, but not economical or practical. A wire mesh might also be adequate and much cheaper and more practical from the standpoint of the circulation of air.

Protective Mesh — Transmission through a
Grid of Wires

How effective is a wire screen? It is apparent that the more wide open the mesh, the less the shielding. In order to answer this question, some early published work (3) was reviewed. Laboratory tests were made both at the Bell Telephone Laboratories in Whippany, N. J., and at Wheeler Laboratories in Great Neck, N. Y., in an attempt to check the earlier work. It was found that *good* agreement was obtained at the *same frequency* as that of the earlier experiment, but that quite *poor* agreement was obtained at *three times that frequency*.

Looking further into the published literature, the formulas of Schelkunoff and Sharpless (4) and that of Marcuvitz (5) were tried. A somewhat better agreement with the data was obtained but still neither formula was satisfactory. However, by combining the two formulae properly, since one neglected one thing and the other neglected something else, an empirical formula was derived which checked all of the available experimental data quite well. This empirical formula was then revised so as to make it suitable for use in nomographic form and the nomograph was constructed, as seen in Figure 7. It applies to normal incidence on a grid of parallel wires of diameter $2r$ having a spacing, a, between centers. The electric vector is assumed to be parallel to the wires. To use the nomograph, a straightedge is aligned with one point at the left corresponding to the spacing in wavelengths and another point at the

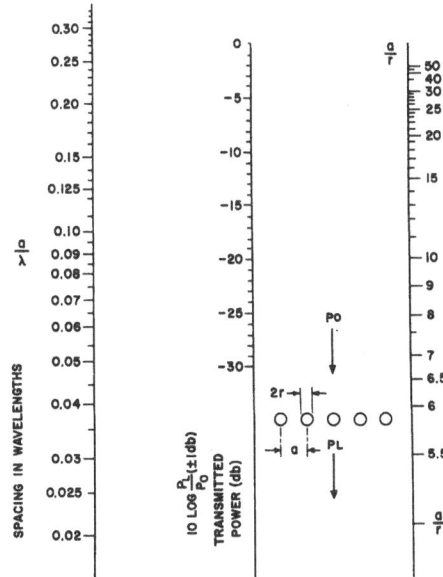

FIG. 7. Transmission through a grid of wires of radius r and spacing a.

right corresponding to the ratio of the spacing to the radius of the wire. At the point where the straightedge crosses the line in the middle labeled "Transmitted Power in db," the shielding effectiveness of the grid of wires is expressed in decibels.

The modified empirical formula upon which this nomograph was constructed is:

$$P_0/P_L = B^2/4 \tag{32}$$

where
$$B = \frac{\lambda}{a} \frac{1}{\ln\left(\dfrac{0.83 \exp(2\pi r/a)}{\exp(2\pi r/a) - 1}\right)}$$

$P_0 =$ incident power
$P_L =$ transmitted power

The accuracy of the nomograph appears to be slightly better than ± 1 db, judging from a comparison with the available measured data.

The results are applicable equally well to a screen of perpendicular wires by ignoring one or the other set of parallel wires forming the mesh.

ACKNOWLEDGMENT

The authors are grateful for the many helpful suggestions of their colleagues within the Bell System and for their cogent observations and comments acquired through frank discussions and internal memoranda. The following contributors have participated: Messrs. C. E. Becraft, N. H. Brown, A. B. Crawford, W. C. Jakes, T. W. Madigan, J. R. Ranucci, G. M. Smith, and Miss M. G. Williams.

The motivation for this review arose originally in connection with the research and development program of the NIKE guided missile systems. This phase of the study was sponsored by the U. S. Army Ordnance Corps. under contract with Western Electric Company (#DA-30-069-ORD-1955). The liaison personnel, Col. R. C. Miles, Major R. D. Fuller, and Mr. J. J. Turner, stimulated this effort with many fruitful discussions.

References

1. *Radio Frequency Hazards,* Bu Aer or USAF T.O. 31-1-80, 15 April 1958.

2. Bickmore, R. W., and Hansen, R. C., "Antenna Power Density in the Fresnel Region," *Proc. I.R.E., 47,* 2119 (Dec. 1959).

3. Hayes, W. D., *Gratings and Screens as Microwave Reflectors,* M.I.T. Rad. Lab. Report No. 54-20, April 1, 1943 (now declassified).

4. Schelkunoff, S. A., and Sharpless, W. M., *Reflecting Systems for Antennas,* U. S. Patent 2,292,342, Aug. 4, 1942.

5. Marcuvitz, N., *Waveguide Handbook,* M.I.T. Rad. Lab. Series, Vol. 10, McGraw-Hill, New York, 1951.

6. Jahnke, E., and Emde, F., *Table of Functions,* Dover Pub., New York, 1943, p. 34.

Development of a Garment for Protection of Personnel Working in High-Power RF Environments

MARTIN R. REYNOLDS
Filtron Company, Inc.

1.0. INTRODUCTION

A REQUIREMENT of high-power radars similar to those in the Ballistic Missile Early Warning System (BMEWS) is 24-hr operation. Inoperative periods for maintenance, inspection, and repairs must be kept to an absolute minimum. It is therefore likely that many maintenance functions will be performed in outdoor areas exposed to dangerously high field intensities. A tentative, safe power-density exposure level for human beings has been set at 10 mw/cm². Calculated average power-density levels at BMEWS sites in some areas are higher than 20 times this tentative value (1).

The major requirements of an effective protection garment are:

1. Reduction of high-level fields external to the garment to safe levels inside the garment.

2. Protection against high-voltage gradients built up on the garment.

3. Minimum restriction of visibility, mobility, and dexterity.

The feasibility of an rf suit had been investigated earlier by Filtron; the present program is a continuation of that work (2).

2.0. ELECTRICAL PROBLEMS

2.1. Shielding

Before we could select and test possible shielding materials, it was necessary to set a minimum requirement on the attenuation desired.

71

From a knowledge of the expected average power-density levels at high-power radar sites, it was possible to select a definite lower limit for acceptable shielding efficiency. A minimum shielding attenuation of 40 db, which includes a safety factor, was decided upon. Shielding can be obtained either by absorption or reflection of the rf energy or a combination of these. A little thought will show that the absorption technique demands a suit which would be too cumbersome to be worn and would also require a means to dissipate the heat energy absorbed. The most reasonable technique is to reflect the energy. This might be accomplished by wearing a solid, metallic enclosure, like the medieval suit of armor. A better enclosure would be one made from cloth which has a continuous metallic coating on its surface. The amount of reflective shielding which such a material produces is a function of its intrinsic impedance as compared to the impedance of the impinging rf field. The greater the difference between these impedances, the more rf energy will be reflected. To be on the safe side in our measurement of shielding efficiency, a low-impedance field, as opposed to a plane wave, should be used. This can be obtained in the laboratory by using a loop. The wave impedance close to the loop is very small and increases with distance from the loop, eventually becoming 377 ohms at great distances with respect to a wavelength. Using the field adjacent to the loop will produce the most rigorous test of a material's shielding effectiveness.

Fig. 1. Material shielding effectiveness test setup.

FIG. 2. Close-up of loops and shielded test box.

Another important factor is the shielding enclosure. It is not enough just to provide a metallic surface between the radiating source and the person to be protected. The shielding surface must enclose the wearer's entire body from head to toe and fingertip to fingertip. Moreover, the shield must be continuous. There must not be any major openings as for the head, hands, and feet in the manner of normal clothing.

The laboratory test setup is illustrated in Figure 1; a close-up of the shielded test box and loops is shown in Figure 2. It should be noted that the upper loop is shown too high. During measurement, the loops are separated only by the thickness of a dielectric plate.

The cloth to be measured is placed between the loops so that it completely covers and makes good contact with all edges of the box's rectangular opening as shown in Figure 3. The box is now a shielded enclosure for the inside loop. All outside rf energy must enter through the shielding cloth.

The difference in receiver readings without the cloth and then with the cloth is the shielding of the cloth in decibels.

2.1.1. MATERIALS

Many types of metalized cloth were prepared and tested. After evaluation for several factors, only three of these provided enough

Fig. 3. Shielding efficiency measurement of cloth sample.

attenuation in the frequency band of interest; these were light metalized nylon, heavy metalized nylon, and metalized heavy marquisette known as Attenutex. Curves of attenuation versus frequency for these are shown in Figure 4. The specially metalized heavy marquisette is seen to be superior to the other materials, and more detailed measurements were made of its characteristics. Shielding measurements were conducted from 0.1 mc to 1000 mc on one, two, and three layers of the Attenutex. The results are shown in Figure 5. From the curves, the attenuation for the double layer is seen to be over 40 db from 100 to 1000 mc.

2.1.2. Seams

The next step was to produce a seam which was mechanically strong and yet preserves electrical continuity between the joined cloth panels. After considerable experimentation, it was found that a double-needle seam, stitched to specifications, preserved the shielding integrity of panels of cloth sewn together. The test box (see Fig. 2) was used to check the attenuation of different seams against that of panels of whole cloth. It was found that the selected seam produced about the same shielding as a single, uncut layer of cloth (see Fig. 6). It is interesting to compare the increased electrical

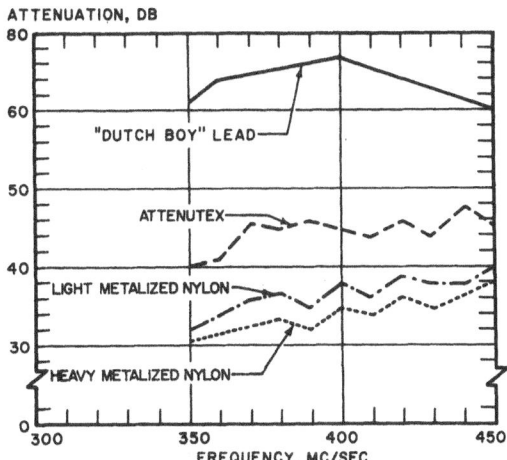

FIG. 4.   Attenuation vs. frequency for several types of
metalized cloths.

effectiveness of a double-stitched seam over that of a single-stitched
seam.   The material was a different type of nylon netting.   (See
Fig. 7.)   The double seam produced 5 to 15 db more attenuation
than a single-stitched seam.

2.1.3.   SUIT ENTRANCE OPENING

This suit entrance opening must be sealed against rf radiation
and still remain flexible enough for body movement.   It must also

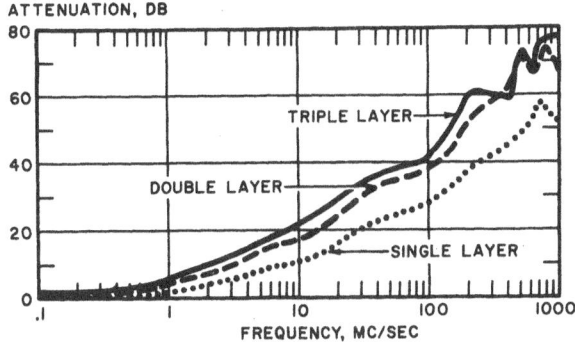

FIG. 5.   Shielding effectiveness of one, two, and three layers of
Attenutex.

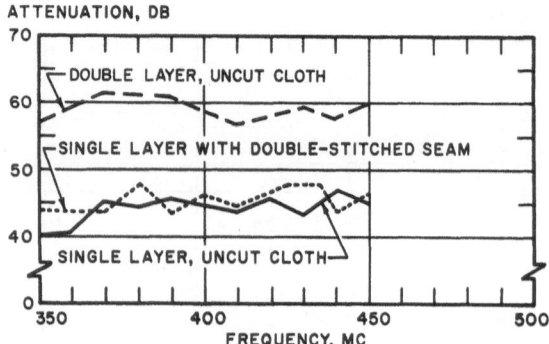

FIG. 6. Shielding effectiveness of Attenutex with and without seams.

withstand repeated opening and closing without degradation of its shielding effectiveness. Two possibilities were investigated.

The first was a two-part suit. The upper section would be similar to a hooded parka, while the lower one would be similar to a pair of high-waisted overalls. The metalized cloth panels of these two sections would make electrical contact throughout an area over the body extending from beneath the arms down to just below the waist-line. The major disadvantage of a two-part suit is ensuring the complete metalized cloth-to-cloth contact necessary for uninterrupted current flow. It was felt that this continuous surface contact between overlapping metalized cloths would be almost impossible to achieve.

The only suit configuration that seemed to be practical was a one-piece garment with a single entrance opening. This opening must remain sealed while the garment is being worn and further must be

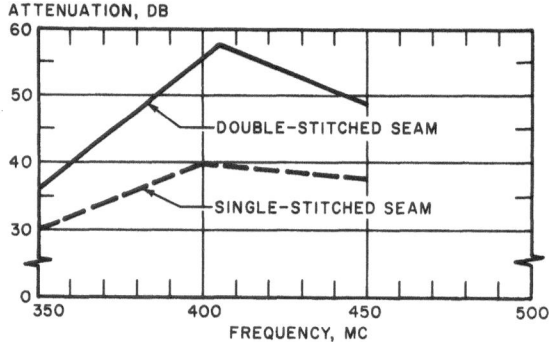

FIG. 7. Shielding effectiveness of double and single seams in nylon net material.

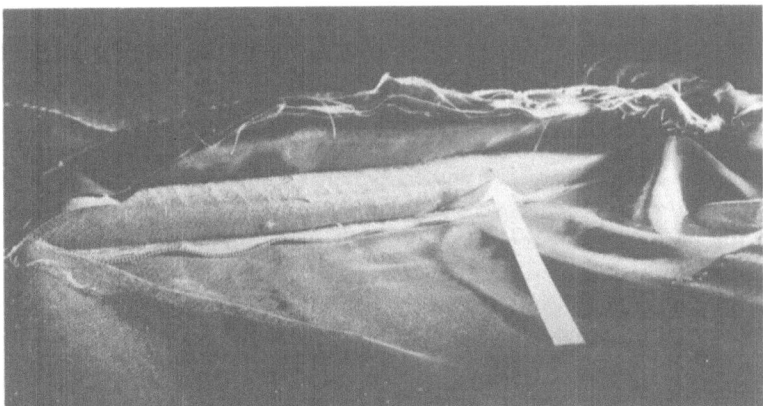

Fɪɢ. 8. Closure after severe abrasion of one surface.

opened easily in order to remove the suit. An ordinary metal zipper with metalized tapes seemed to be the answer. However, it was found that the metal teeth of any zipper are always attached to a cotton beading, which is usually sewed on to tapes of nylon. The nylon tapes could be metalized with no trouble, but the cotton beading could not be metalized because it was too absorbent. Experimentation, with the cooperation of several zipper manufacturers, showed that a beading made of nylon would not permit the zipper teeth to be crimped on with present machinery. The possibility of depending upon a zipper for electrical continuity was abandoned.

The closure used in the rf suit which is presently being prepared for patent application provides a contiguous surface contact between the two mating cloths and is completely flexible with body movement. Electrical shielding tests on the closure itself showed at least 60 db of attenuation. Even after severe sandpaper abrasion of one of the mating surfaces, the shielding was more than adequate. An abraded sample is shown in Figure 8.

A closure location at the rear of the suit was chosen because it allows entry through an opening of minimum length. An opening in the front or the side would have to be much longer and might interfere with mobility.

2.2. Voltage Stresses

In the high-power-density areas contemplated, electric field intensities of several thousand volts per meter exist in free space.

Since the rf suit uses metalized fabric its conductivity is considerably lower than that of solid metal, and it is likely that high rf voltages can be developed between different parts of the garment. Even if the suit were an excellent conductor, it is possible that the wearer might place his body between two otherwise disconnected metallic members being irradiated by the rf field. An example might be the tightening of an overhead pipe while standing on a metal platform. If the platform and pipe are not otherwise bonded together through a lower impedance path at the radio frequency, a severe shock can be transmitted through the suit. For these reasons, it is desirable to cover the entire garment with a dielectric insulating material of some kind.

High-voltage-breakdown tests were made on more than 50 materials — some porous, some impervious. It was decided to use an impervious specially constructed Neoprene-coated nylon cloth as the outer shell of the suit. Although only 0.009 in. thick, direct voltage tests have shown this fabric to protect against potentials higher than the 4000-v design objective. With this impervious shell, fresh-air intake to the suit is through a screened eye visor, which is the only opening when the suit entrance is closed. This suit was designed for low-temperature environment and the almost total body inclosure is not restrictive in this case.

3.0. SENSORY PROBLEMS

3.1. Visibility

Visibility is provided through a screened semicircular visor in front of the eyes. Other materials, such as thin metal films on glass, were investigated, but their electrical attenuation was much lower than screening, with not much more visibility. Eyepieces designed on the basis of waveguide-below-cutoff apertures were considered also. These transmit light with no loss, but would seriously restrict the field of view.

Shielding efficiency measurements were made on wire-mesh screening of several size apertures. The results for three of these are shown in Figure 9. It can be seen that meshes finer than about 18 strands/in. provide over 40 db of attenuation above 100 mc. The screening selected for the suit is a 24-strand/in. 0.014-in. copper mesh. The added radiation attenuation is considered worthwhile for protection of the eyes.

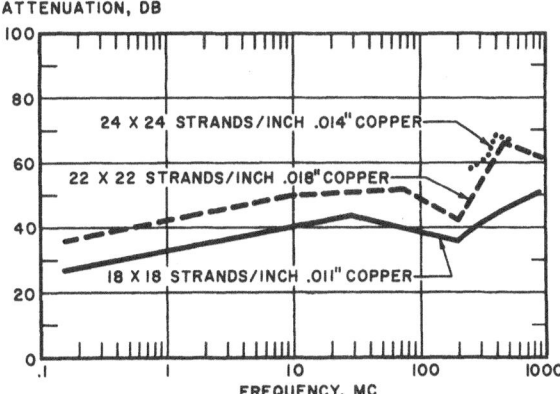

FIG. 9. Shielding effectiveness of wire screenings in a low-
impedance field.

Electrical continuity between the layers of shielding cloth and
metallic screening is provided by rectangular metal frames and a
Neoprene gasket. An experimental version of the assembly is illustrated in Figure 10. Pressure between the screening frame and the
cloth is maintained by machine screws spaced approximately 1 in.
on centers. The holes in the inner frame have counterbores milled

FIG. 10. Experimental version of screened eye-visor assembly.

onto the inner concave surface of the frame. The counterbores are used to receive washers, which seal the holes against radiation leakage. The screws and holes are actually coaxial-transmission lines, which can conduct radiation through the entire mask. The washers eliminate this hazard. The holes in the shielding cloth are made slightly under the screw diameter size so as to make continuous contact around each screw. This also reduces the possibility of a coaxial-transmission mode through the visor.

3.2. Ventilation and Perspiration

The requirement for an impervious dielectric outer covering has presented severe limitations on ventilation and perspiration. The suit was originally intended for use at low temperatures and this model is not considered usable for moderate climates in its present form. The exact temperature limits or range within which its wearing is comfortable have not yet been determined, but low-temperature tests have been made at a constant temperature of $-10°F$ ($-23°C$).

3.3. Mobility and Dexterity

The suit has been generously proportioned to provide maximum mobility. It is possible to run, squat, bend, and twist without restriction.

Dexterity has been compromised somewhat because of the decision to use a simple mitten construction for the hands. To provide a fingered hand would be difficult because of the several layers of different materials required. These would make individual fingers too bulky and clumsy, and would defeat the purpose of dexterity. The present mitten design permits moderately fine operations, such as manipulation of instrument dials and ordinary hand tools.

4.0. ENVIRONMENTAL AND MECHANICAL PROBLEMS

4.1. Low-Temperature Environment

All the materials considered for the rf suit were tested in a cold chamber through the temperature range from 0 to $-60°C$. A partial tabulation of results is given in Table I. The criterion of a suita-

TABLE I

Low-Temperature Tests of R-F Suit Components

Material	Temperatures in Degrees Centigrade and Fahrenheit						
	0°C	−10°C	−20°C	−30°C	−40°C	−50°C	−60°C
	32°F	14°F	−4°F	−22°F	−40°F	−58°F	−76°F
Neoprene Foam Rubber	Soft	Soft, springs back slowly	Pliable, springs back slowly	Hard, deformable	Hard, nondeformable	Hard, nondeformable	Hard, nondeformable
White Foam Rubber	Soft	Soft	Soft, springs back slowly	Pliable, springs back slowly	Hard, deformable with effort	Hard, nondeformable	Hard, nondeformable
Foam Polyurethane 765, Charcoal	Soft	Soft	Soft	Soft, springs back slowly	Pliable with effort	Hard, deformable with effort	Hard, nondeformable
Foam Polyurethane 725, Green	Soft	Soft	Soft	Pliable	Pliable	Crisplike hardness, pliable	Brittle-like hardness, thaws with compression
Foam Polyurethane 805, White	Soft	Soft	Soft	Soft	Pliable	Crisplike hardness, pliable	Crisplike hardness, thaws with compression
Foam Polyurethane 800, Green	Soft	Soft	Soft	Soft	Pliable	Crisplike hardness, easily pliable	Crisplike hardness, easily pliable, thaws with compression
Neoprene Sheet Rubber	Soft	Soft	Soft	Soft	Stiff, but flexible	Stiff, but bendable	Stiff, but bendable
Neoprene Coated Nylon	Soft	Soft	Soft	Soft	Soft	Soft	Stiff, but easily pliable

NOTE: No changes observed with heavy marquisette shielding cloth or Talon "big zip" zipper.

ble material was its pliability at extreme low temperatures. At 10° intervals, the materials were compressed or folded by hand and visual observations were made.

The Neoprene sheet rubber is a gasketing material that is used in the eye visor. It was found to remain resilient enough so that the metallic cloth was in pressurized contact with the screen frame down to $-60°C$. The Neoprene-coated nylon, used for the suit's outer shell, was practically unchanged in its pliability throughout the temperature range.

The Attenutex cloth and the Talon zippers were also unchanged at low temperatures.

Results on all the materials used in the present suit indicate that the suit will remain flexible down to at least $-60°C$.

4.2. Abrasion

Mechanical abrasion and wrinkling tests were necessary to check the permanency of the material's shielding effectiveness. A real danger would exist if the metallic coating were to come off unnoticed after a few wearings.

After this test, the cloth was subjected to abrasion. Two squares of cloth were each attached to flat blocks of wood. The blocks were then rubbed vigorously together, abrading the cloth surfaces. Shielding effectiveness was measured after each of six 5-min intervals. No significant degradation of shielding effectiveness was found.

Because it is important for the closure to maintain good surface contact between mating cloths, a sandpaper abrasion test was made. A full-length closure was placed atop a large, screened box. Each half of the closure had two layers of Attenutex cloth, as in the present suit. The lower mating surface was abraded with fine sandpaper for 1 hr. The effect on the cloth is shown in Figure 8. The shielding efficiency remained over 50 db even after the outer cloth of the lower surface was ripped.

5.0. SUIT CONSTRUCTION

The present suit consists of several layers of material. The outer shell is a Neoprene-coated nylon for electrical insulation and abrasion protection. The next two layers are Attenutex cloth for shielding. A third layer of this cloth has been sewn under and around the feet for added strength.   The innermost lining is a strong, cotton broad-cloth.   Each layer is a separate garment, and all are tacked together for strength at specific points.   The rf suit is illustrated in Figure 11; it weighs just under 10 lb without the boots.   In warmer weather, the arctic boots shown need not be worn.   In extremely cold weather, a parka and pants can be worn over the suit.

6.0.   SUIT TESTS

The shielding effectiveness of a complete suit has been checked at low power levels.   The results show that at no location was there less than 45 db shielding, and at 9 locations greater than 50 db shielding was obtained.   It should be remembered that this was a low-impedance field measurement and the attenuation of plane waves is expected to be even greater.

Fig. 11.   Radio frequency protective suit.

7.0. CONCLUSIONS

A planned program of design and development has led to the manufacture of an rf protective suit. It can be worn with safety in an electromagnetic field whose power density is ten thousand times more intense than the present safe limit. The suit allows complete freedom of movement with unobstructed visibility. It was designed to withstand severe mechanical and environmental tests. These have shown the suit to be extremely rugged and capable of maintaining its protective function without restricting mobility and dexterity.

References

1. Filtron Company, Inc., *Power Density Levels in the Region Between the BMEWS Detection Radar Reflector and Scanner Building,* Report No. TM-1063-14, 10 November 58 (Secret).

2. Harley, J., *Garments for RF Protection of Personnel in High-Power-Density Environments,* Filtron Company, Inc., Report No. TM-1063-22, 13 March 1959 (Secret).

The Time Constants of Pearl-Chain Formation*

M. Saito and H. P. Schwan

Electromedical Laboratory
The Moore School of Electrical Engineering
University of Pennsylvania
Philadelphia, Pennsylvania

INTRODUCTION

IN SPITE OF THE FACT that the phenomenon of pearl-chain formation has been drawing the attention of many biophysicists, no quantitative analysis has been conducted on this phenomenon except that of Krasny-Ergen (1). His analysis invites the following comments and criticism:

1. Only the perfectly conducting spherical particles are considered, a case of little biological interest.

2. It is not completely clear whether we can equate the decrease of the potential energy to kT at the threshold of pearl-chain formation.

3. No relation is obtained between the pearl-chain formation threshold and the particle concentration.

A more general investigation has been conducted by the authors to remove all of these restrictions (2, 3). The model used in the investigation is shown in Figure 1. Two dielectric spheres of complex dielectric constant $\in_1^*$ are imbedded in a homogeneous medium of complex dielectric constant $\in_2^*$ and are exposed to a uniform field E_0. The two spheres are not allowed to be separated by more than

* This work has been supported by the contract between The Office of Naval Research and The Moore School of Electrical Engineering, University of Pennsylvania.

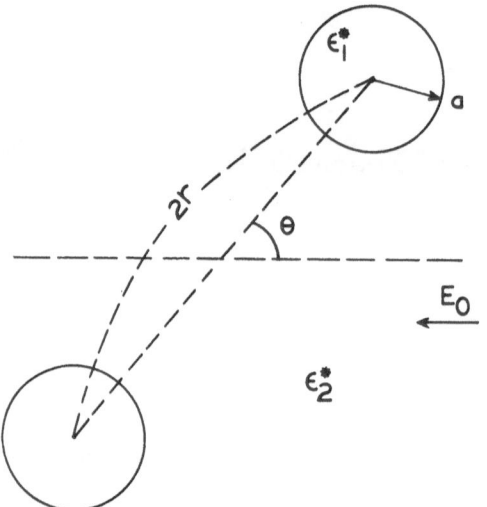

Fig. 1.

a distance $2R$, which is related to the concentration of the suspension by

$$\text{concentration} = 74\% \times (a/R)^3 \qquad (1)$$

assuming the case of a uniform density distribution of particles. This assumption appears only at the first moment artificial, since in actual cases one sphere can always find its pair within a certain distance.

The change in potential energy due to the interaction of the spheres was calculated in two ways: (1) By neglecting the influence of one sphere on the induced charge of the other sphere (dipole approximation), and (2) by solving the field distribution exactly.

Then the mean distance between the spheres was calculated from the Boltzmann distribution formula. An example of the results is shown in Figure 2, where K_0 is the parameter defined by

$$K_0 = U\big|_{\substack{r=a \\ \theta=0}} \sim E_0^2 \qquad (2)$$

The threshold for the pearl-chain formation K_{th} of K_0 is defined by the point where the mean distance is

$$\bar{r} = (\bar{r}\big|_{E_0=0} + a)/2 \qquad (3)$$

The dipole approximation gives a threshold value K_{th} of about 9. This approximation can be shown to be correct within an error of

50% unless the dielectric constants of the particles and the medium differ by a factor larger than 10. Using this approximation, the threshold field intensity is given by

$$E_{th} = 1.7 \, a^{-3/2} |2 + (\epsilon_1^*/\epsilon_2^*)| \sqrt{\frac{kT}{(\epsilon_1 - \epsilon_2) \, Re[(\epsilon_1^* - \epsilon_2^*)/\epsilon_2^*]}} \quad (4)$$

The details of this analysis will be given elsewhere (2). Here another question remains to be solved: Does the threshold value depend on other parameters than the r.m.s. value of the field intensity?

As far as the steady state analysis summarized above is concerned, the threshold depends only on the r.m.s. field intensity and the properties of the particles and the environment. However, some reports seem to indicate also a dependence on the peak value of the field (4, 5).

In order to clarify this question, the transient behavior of the particles engaged in the process of the pearl-chain formation must be considered. This paper summarizes the results of our investigations of the speed with which a pearl-chain formation takes place under various circumstances.

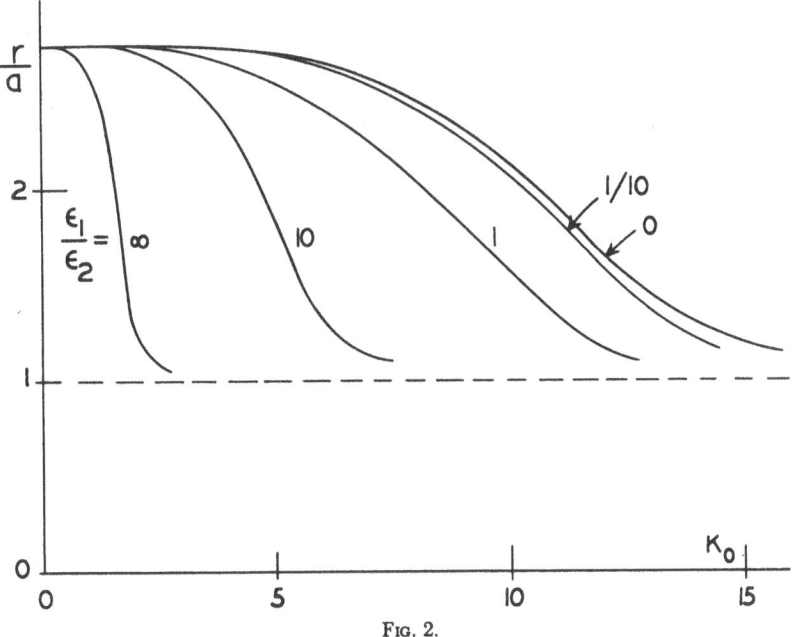

Fig. 2.

FUNDAMENTAL RELATIONS

The probability density function ρ that the particle locates itself
at a particular position in space at a given instant is now a function
of r, θ, and t, though it was a function of only r and θ when considering the steady state. The change of ρ in time and in space is determined by the Smoluchowski equation:

$$(\partial\rho/\partial t) = D\Delta\rho + (1/2f) \text{ grad } U \cdot \text{grad } \rho \tag{5}$$

and
$$D(\partial\rho/\partial n) + (\rho/2f)(\partial U/\partial n) = 0 \tag{6}$$

at the boundary of the region shown in Figure 3,

where
$$D = (kT)/(2f) \tag{7}$$

$$f = 6\pi\eta a \tag{8}$$

and using the dipole approximation,

$$U = -K_0\, kT(a/r)^3\, P_2(\cos\theta) \tag{9}$$

where P_2 is the Legendre polynomial of second degree. The region
where this equation is to be considered is shown in Figure 3. The
factor of 2 appearing in eqs. (5), (6) and (7) differs from the ordinary

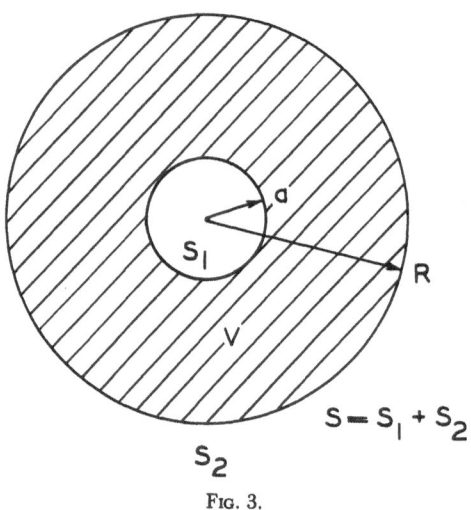

FIG. 3.

Smoluchowski equation. It is due to the particular coordinate system used in this paper.

Assuming a distribution function that changes with time exponentially,

$$\rho \sim e^{-\nu t} \tag{10}$$

eq. (5) reduces to

$$\Delta\rho + (1/kT)\,\text{grad } U \cdot \text{grad } \rho + \lambda\rho = 0 \tag{11}$$

where

$$\lambda = \nu/D \tag{12}$$

Equations (11) and (6) establish an eigenvalue problem, and the eigenvalue λ is related to the inverse of the time constant ν by eq. (12).

TIME CONSTANTS FOR ZERO FIELD INTENSITY

For zero field intensity, eqs. (11) and (6) reduce to

$$\Delta\rho + \lambda\rho = 0 \tag{13}$$

and

$$\partial\rho/\partial n = 0 \tag{14}$$

at the boundary. The solutions of this problem are

$$\rho_{nm} = \{C_{1nm}j_n(c_{nm}r) + C_{2nm}n_n(c_{nm}r)\}\,P_n(\cos\theta) \tag{15}$$

$$\lambda_{nm} = c_{nm}^2 \tag{16}$$

where j_n and n_n are spherical Bessel functions of the first and second kind, respectively, and P_n is the Legendre polynomial of nth degree. C_{1nm}, C_{2nm}, and c_{nm} are determined from

$$C_{1nm}j_n'(c_{nm}a) + C_{2nm}n_n'(c_{nm}a) = 0 \tag{17}$$

$$C_{2nm}j_n'(c_{nm}R) + C_{2nm}n_n'(c_{nm}R) = 0 \tag{18}$$

The inverse of the time constant is given by

$$\nu = x_{nm}^2 kT/12\pi\eta a^3 \tag{19}$$

where x_{nm} is the mth root of

$$j_n'(x)\,n_n'(Rx/a) - j_n'(Rx/a)\,n_n'(x) = 0 \tag{20}$$

It is to be noted that the time constants are proportional to the cube of the radius.

For $R/a = 3.6$, which corresponds to the concentration of about 1.6%, the values of x_{nm} are found to be

$$x_{01} = 1.37 \qquad \cdots$$

$$x_{21} = 0.92 \qquad x_{22} = 1.97 \qquad \cdots$$

Because of the symmetry of our model, only the modes with even n are significant. The modes which take large values at $r = a$ and $\theta = 0$ may be significant in the pearl-chain formation. In our cases, the mode $n = 2$, $m = 2$ is the one with this property, and the value of 1.97 for x corresponds to a time constant of 2.2 sec for $a = 1\mu$ and at room temperature.

EFFECT OF FIELD INTENSITY ON TIME CONSTANTS

As eqs. (11) and (6) contain the terms involving the field strength, the eigenvalues and consequently the time constants are functions of the field intensity. There are several possibilities to calculate the effect of the field intensity on the time constants:

1. Perturbation Method

Equations (11) and (6) are rewritten in the form

$$\Delta\rho + \delta L[\rho] + \lambda\rho = 0 \tag{21}$$

$$(\partial\rho/\partial n) + \delta f(r,\theta)\rho = 0 \qquad (r = a,R) \tag{22}$$

where $$\delta = 3K_0 a^3 \tag{23}$$

$$L[\rho] = (1/r^4)\, P_2 \cos\theta(\partial\rho/\partial r) + (1/r^5)\sin\theta\cos\theta(\partial\rho/\partial\theta) \tag{24}$$

$$f(r,\theta) = -(1/r^4)\, P_2 \cos\theta \qquad (r = a) \tag{25}$$

$$= +(1/r^4)\, P_2 \cos\theta \qquad (r = R) \tag{26}$$

Assume that the solution of these equations can be expanded into the power series of δ for small values of δ:

$$\rho = \rho_0 + \delta\rho_1 + \delta^2\rho_2 \cdots \tag{27}$$

$$\lambda = \lambda_0 + \delta\lambda_1 + \delta^2\lambda_2 \cdots \tag{28}$$

where ρ_0 and λ_0 are the solutions of the problem for zero field intensity. Substitution of eqs. (27) and (28) into eqs. (21) and (22), and further calculations yield

$$\lambda_1 = (-\int_v \rho_0 L[\rho_0]dV + \int_s f\rho_0^2 \, dS) \Big/ \int_v \rho_0^2 \, dV \qquad (29)$$

The regions where the integrals must be evaluated are shown in Figure 3. The evaluation of the integrals appearing in this formula results in

$$\lambda_1 = \xi c_{nm}^5 I_{2nn} \qquad (30)$$

where (6)

$$\xi = \frac{-\int_{C_{nm}a}^{C_{nm}R} \frac{z_n z_n'}{x^2} \, dx + \int_{C_{nm}a}^{C_{nm}R} \frac{z_n^2}{x^3} \, dx + \left[\frac{z_n^2}{x^2}\right]_{C_{nm}a}^{C_{nm}R}}{\frac{1}{2n+1}\left[x^3\left\{1 - \frac{n(n+1)}{x^2}\right\}z_n^2\right]_{C_{nm}a}^{C_{nm}R}} \qquad (31)$$

$$z_n = C_{1nm}j_n(x) + C_{2nm}n_n(x) \qquad (32)$$

$$I_{pqr} = \int_{-1}^{1} P_p(\mu)P_q(\mu)P_r(\mu)d\mu \qquad (33)$$

The integrals remaining in eq. (31) must be evaluated numerically. Using the value of ξ thus obtained,

$$(\delta\lambda_1)/(\lambda_0 K_0) = (\% \text{ change in } \lambda)/(\text{unit increase in } K_0)$$

$$= \xi I_{2nn} c_{nm}^3 a^3 \qquad (34)$$

Taking an example of $R/a = 3.6$,

$$(\delta\lambda_1)/(\lambda_0 K_0) = 0 \qquad \text{for } n = 0$$

$$= -0.02\% \qquad \text{for } n = 2, m = 1$$

$$= -0.84\% \qquad \text{for } n = 2, m = 2$$

Therefore, it can safely be stated that the time constants are not strongly dependent on the field intensity when the latter is small.

2. Difference Method

The perturbation method is restricted to the cases of weak fields. Better techniques for the cases of strong fields are the difference method and the variational method.

In the difference method (7), the partial differential equation is converted into a difference equation, which is solved successfully by a digital computer. This method has not been used by the authors, since the goal of the present work is to obtain some estimates on the speed of formation of the pearl-chain, and not to give a precise description of the transient states.

3. Variational Method (8)

Changing the variable ρ by means of

$$\rho = P \exp \{ -1/(2kT) \} \tag{35}$$

eqs. (11) and (6) are transformed into

$$\Delta P - (1/4k^2T^2)|\operatorname{grad} U|^2 P + \lambda P = 0 \tag{36}$$

$$(\partial P/\partial n) + (P/2kT)(\partial U/\partial n) = 0 \tag{37}$$

The variational principle shows that the nth eigenvalue of this eigenvalue problem is characterized by

$$\lambda_n = \operatorname*{Max}_{\psi_i} \operatorname*{Min}_{P} (J[P]/K[P]) \tag{38}$$

with the conditions

$$\int_v P\psi_i dv = 0 \qquad (i = 1,2,\ldots n-1) \tag{39}$$

where

$$J[P] = \int_v [|\operatorname{grad} P|^2 + (1/4k^2T^2)|\operatorname{grad} U|^2 P^2]dV$$
$$+ \int_s (1/2kT)(\partial U/\partial n) P^2 dS \tag{40}$$

$$K[P] = \int_v P^2 dV \tag{41}$$

The regions where the integrals are to be carried out are shown in Figure 3. The meaning of eqs. (38) and (39) is that the minimum of J/K is obtained for fixed ψ_i's and then the maximum of these minimum values is obtained by changing ψ_i (maximum-minimum principle).

In order to solve numerically the variational problem represented by eqs. (38) to (41), a linear expansion for the solution is assumed:

$$P = \sum_{i=1}^{n} a_i v_i(r, \theta) \tag{42}$$

Substitution of eq. (42) into eqs. (40) and (41) gives

$$J[P] = \sum_{i,j=1}^{n} A_{ij} a_i a_j \tag{43}$$

$$K[P] = \sum_{i,j=1}^{n} B_{ij} a_i a_j \tag{44}$$

From these, the characteristic equation is obtained:

$$\det[A_{ij} - \lambda' B_{ij}] = 0 \tag{45}$$

The roots λ_i' of this equation give the upper bounds to the true eigenvalues λ_i. For the details of this method, see, for example, Collatz (8).

In our calculations, the bases v_i are chosen as

$$v_{nm} = r^m P_n \cos \theta \tag{46}$$

and only four v_{nm}'s are determined since the extent of the numerical calculations increases rapidly with the number of v_{nm}'s. The four v_{nm}'s correspond to

$$(n,m) = (0,0), (0,1), (2,0), (2,1)$$

In order to obtain some idea as to the accuracy of this approximation, the results of the calculation for zero field intensity by this method and by the exact method given in Part 1 of this section are compared in Table I. It seems that the results of this method are correct within a factor of 2 or so.

Numerical calculations are carried out for $a = 1$ and $R = 3.6$, and the results are shown in Figure 4. It may be useful to note that for $a = 1\mu$ and at room temperature, $\lambda = 1$ corresponds to a time constant of 9 sec and that the time constants are proportional to the

TABLE I

Values calculated by approximate method	Values calculated by exact method
0	0
0.878	0.846
2.99	1.88
5.56	3.88

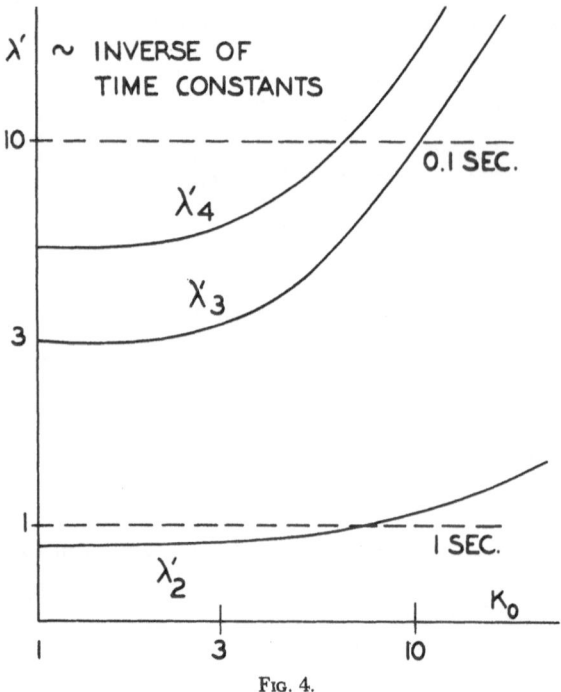

FIG. 4.

cube of the radius. In Figure 4, λ_1 is always zero, and λ_1' does not appear; for strong fields other λ_i's are expected to be proportional to K_0 or the square of the field intensity, though λ_3' and λ_4' increase more rapidly in the figure owing to the poor approximation.

THE CASE OF PULSED FIELD

Consider the case of the pulsed applied field shown in Figure 5, and let τ_p, τ_a be the time constants of a mode corresponding to the power P_p, P_a, respectively. Obviously,

$$P_a = P_p t_1 / t_0 \qquad (47)$$

and as is pointed out in the preceding section, the time constants are inversely proportional to K_0 or power of the applied field:

$$\tau_a / \tau_p = P_p / P_a \qquad (48)$$

Consequently,

$$\tau_a / t_0 = \tau_p / t_1 \qquad (49)$$

This equation shows that, if t_0 is much smaller than τ_a, t_1 is also much smaller than τ_p, and the distribution function does not change appreciably during the period of the applied field.

On the other hand, it can be shown that the time mean of the distribution function is almost completely determined by the r.m.s. value of the field intensity, if the distribution function does not change appreciably during the period (3). Therefore we can conclude that, if t_0 is much smaller than the time constant corresponding to the average power, the time mean of the distribution function is almost completely determined by the r.m.s. value of the applied field.

In ordinary radar systems, t_0 is of the order of 1msec, while τ_a is of the order of 1 sec. So only the r.m.s values are effective in the pearl-chain formation.

Next, we compare the tolerance dose rate level for microwave hazards due to heat development with the one for pearl-chain formation. The threshold for the former has been set at 10 mw/cm^2, corresponding to the field intensity of somewhat less than 1 v/cm in muscle and similar soft tissues. The formula for the threshold for the pearl-chain formation as given in another paper by the same authors (2) gives about 80 v/cm for a radius of 1 μ and 2.5 v/cm for a radius of 10 μ. The pearl-chain formation is possible at subthreshold levels only for particles larger than 10 μ, assuming the pulse conditions in the usual radar systems. However, particles of those sizes are of no biological importance since they are not freely movable.

Finally we consider the situation in which the height and duration of the pulse are so chosen as to enhance the pearl-chain forma-

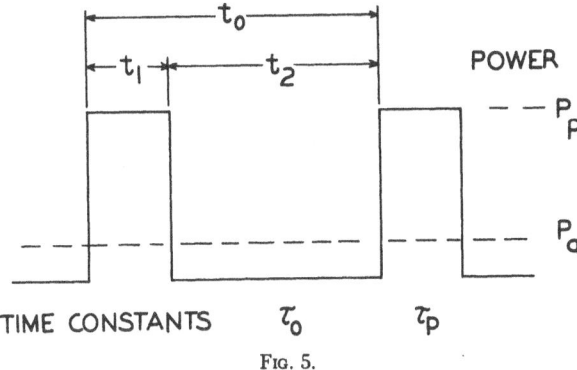

Fig. 5.

tion, and the period is chosen long enough so that the average power can be kept below the 10 mw/cm^2 level. Assuming the pulse height to be equal to the threshold for pearl-chain formation, and the duration of the pulse to be equal to the time constant of the mode $n = 2$, $m = 2$, it is easy to see that t_0 must be larger than about 200 min, independently of the radius as long as it is less than 10 μ. Such values of t_0 are of no practical interest since it is much longer than the thermal time constants involved in heating up the human body or parts of it.

CONCLUSIONS

As the results of the investigations made on the transient behavior of the pearl-chain formation, it is concluded that:

1. The time constants involved in the pearl-chain formation are of the order of a second for the radius of 1 μ, and they are proportional to the cube of the radius.

2. The time constants are not strongly dependent on the field intensity when it is small, and they are inversely proportional to the square of the field intensity when it is large. The transition between the two regions occurs near the threshold.

3. For particles of several microns or of larger sizes, the time constants become as large as hundreds of seconds, and consequently there exists a danger of observing erroneous threshold values in experimental works.

4. The pulsed applied field is as effective as is expected from its r.m.s. value for usual radar systems.

5. The hazards due to heat become pronounced before the field intensity is large enough for the pearl-chain to form if the particle size is less than 10 μ.

Finally we conclude that the phenomenon of pearl-chain formation is of no biological significance in the area of microwave hazards due to radar exposure.

References

1. Krasny-Ergen, W., "Nicht-thermische Wirkungen elektrischer Schwingungen auf Kolloid," *Hochfreq. u. Elektroak.*, *48*, 126 (1936).

2. Saito, M., and Schwan, H. P., "Analysis of a Model for Pearl-Chain Formation," to be published.

3. Saito, M., and Schwan, H. P., "The Transient Behavior of the Pearl-Chain Formation," to be published.

4. Herrick, J. F., "Pearl-Chain Formation," *Proc. 2nd Tri-service Conf. on Biological Effects of Microwave Energy,* 1958, pp. 88–96.

5. Leary, F., "Researching Microwave Health Hazards," *Electronics, 32,* 49 (Feb. 20, 1959).

6. Hobson, E. W., *The Theory of Spherical and Ellipsoidal Harmonics,* Cambridge Univ. Press, 1931, p. 87.

7. Kunz, K. S., *Numerical Analysis,* McGraw-Hill, N. Y., 1957, Chap. 12.

8. Collatz, L., *Eigenwertprobleme,* Chelsea, N. Y., 1948, Chap. 5.

The Effect of Microwave Radiation (24,000 mc) on the Male Endocrine System of the Rat

Samuel A. Gunn, Thelma Clark Gould,
and W. A. D. Anderson
Department of Pathology
University of Miami School of Medicine
Coral Gables, Florida

Our primary purpose in this study was to determine if exposure of the testes of rats to microwaves of 1.25 cm wavelength would produce a functional disturbance in the male endocrine system, even though there might be no morphological evidence of damage to the testes. Imig, Thomson, and Hines (1) studied the effects of 12-cm microwave radiations on the testes, and, although degenerative changes were noted in the germinal epithelium, the Sertoli and interstitial cells apparently remained intact. The indicator of gonadotrophin or androgen output most frequently used in endocrine studies is a measurement of the weight or the histological appearance of the accessory sex organs. It should be emphasized here, however, that there may be functional disturbances in the sex accessory glands which are not reflected in the size or microscopic appearance of these organs (2).

At this time it will be necessary to orient you to the method we have devised to measure variations in androgen output or utilization. In the course of studies on the rat prostate, it has been shown that administered Zn-65 concentrates to a high degree in the dorsolateral lobes of the prostate (3), paralleling the rich natural zinc content of the gland (4).

Figure 1 shows the tissue distribution of Zn-65 at various times following the intracardiac administration of tracer doses of the radioisotope. The graph shows that Zn-65 is concentrated by the dorso-

99

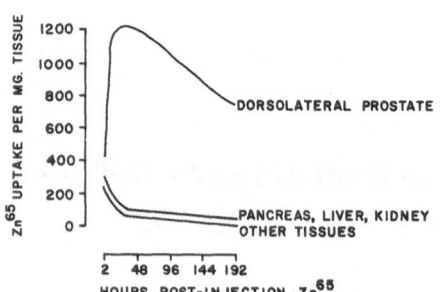

Fɪɢ. 1. Tissue distribution of Zn-65 in the rat various times after intra-cardiac administration of tracer doses of the radioisotope.

lateral prostate and retained while other tissues take up a smaller amount and do not retain it as long (3).

Figure 2 shows that the capacity of the dorsolateral prostate to concentrate Zn-65 is hormonally controlled. The graph indicates that the low level of Zn-65 uptake in the castrate is increased by the administration of testosterone (5). Furthermore, this illustrates that measurement of total Zn-65 uptake is a far more sensitive indicator of androgen level than measurement of the weight of the accessory gland.

Figure 3 shows that the low level of Zn-65 uptake noted in the hypo-physectomized rat is increased by the administration of gonadotrophin (2). Again it is evident that measurement of total Zn-65 uptake is a far more sensitive indicator of gonadotrophin, and consequently of androgen level, than measurement of the weight of the accessory gland (6).

Figure 4 is a simplified diagram of the pituitary-testes-prostate endocrine chain, illustrating the hormonal control of Zn-65 uptake

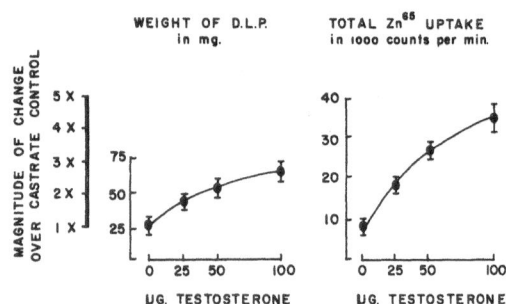

Fɪɢ. 2. Weight and Zn-65 uptake response of the dorsolateral prostate of the castrate rat to various doses of testosterone.

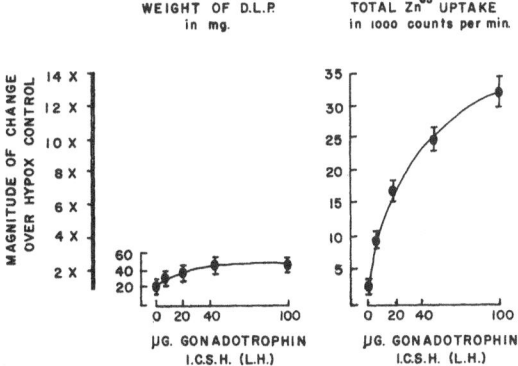

FIG. 3. Weight and Zn-65 uptake response of the dorso-lateral prostate of the hypophysectomized rat to various doses of gonadotrophin.

by the dorsolateral prostate. Note the normal sequence of events. The pituitary secretes gonadotrophin, which is utilized by the testes in the production of testosterone, which is necessary for the dorso-lateral prostate to concentrate Zn-65.

Zn-65 uptake studies have been used in our work to determine whether various forms of radiation will produce a break in this endocrine chain (7, 8). When there is a fall in Zn-65 uptake it indicates that either pituitary, testes, or prostate have been damaged. Testosterone is administered to determine if the end target organ, the dorsolateral prostate, is capable of utilizing this hormone. If this gland

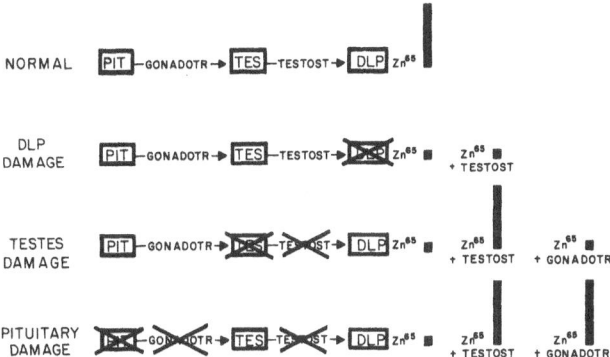

FIG. 4. Simplified diagram of the pituitary-testes-prostate endocrine chain controlling the Zn-65 uptake by the dorsolateral prostate.

does not respond with increased Zn-65 uptake, then one of the defects is in the gland itself. If the dorsolateral prostate does respond to testosterone, denoting that this accessory gland is intact, then it is necessary to rule out testes or pituitary. If the administration of gonadotrophin results in no increase in Zn-65 uptake, the defect is definitely in the testes. If the dorsolateral prostate responds to both testosterone and gonadotrophin, then the fault lies in the pituitary. Now let us turn to some experiments that have been done to illustrate this means of detecting damage to the pituitary-testes-prostate endocrine chain, such as that caused by X-irradiation, for example.

Recent references on radiation biology have emphasized that the spermatogonia are the most sensitive of the testicular elements to ionizing radiation, while the interstitial cells are the most radio-resistant (9–13). Earlier investigators applying total-body X-irradiation (14) or local radiation to the testes of rats pointed out that while the seminiferous tubules were damaged androgen production was not diminished, as evidenced by the appearance of normal size or hypertrophied sex accessory glands.

In our laboratory rats were subjected to 470 r whole-body X-irradiation. At 1-, 2-, and 3-weekly intervals following X-ray exposure, rats were injected with tracer doses of Zn-65, the animals were sacrificed, and the amount of Zn-65 taken up by the dorsolateral prostate was determined. At the same time the testes were examined grossly and microscopically for damage (7).

The microscopic study of the testes of rats sacrificed 1 week following X-irradiation revealed that only 30% of the tubules had spermatogonia. Spermatocytes were normal in distribution and cell population, and the early and late spermatids were normal in appearance and in quantity. The interstitial cells appeared normal. Two weeks following X-irradiation only 10% of the tubules had spermatogonia. Spermatocytes were reduced 20% from normal and were normal appearing. The early and late spermatid population was reduced approximately 30% in many tubules, indicating an arrest in the intensity and rate of spermatogenesis. The interstitial cells appeared normal. Three weeks post-irradiation a progressive loss of spermatogonia was evident, only 5% of the tubules having this layer present. At this time the spermatocytes were reduced 80% from normal. The early and late spermatid population was normal, indicating that the arrest noted at 2 weeks post-exposure was only temporary. The interstitial cells appeared normal. The microscopic picture of testes of rats sacrificed at 2 weeks post-irradiation but treated with testosterone or gonadotrophin showed spermatogonia and spermatocyte population the same as in the untreated rats sacrificed at 2 weeks post-irradiation. However, in the hormone-treated rats the early and late spermatid population was maintained at normal levels, indicating the importance of androgen for the normal progression of spermatogenesis.

The results of the Zn-65 uptake studies on X-irradiated rats are shown in Figure 5. The graphs show that at 6 days post-irradiation the weight and Zn-65 uptake of the dorsolateral prostate was slightly increased (although not significantly different from control values). Thirteen days post-exposure to X-rays there was a 29% fall in the weight of the dorsolateral prostate ($P < 0.01$) and a 36% diminution in the total amount of Zn-65 concentrated by the gland ($P < 0.001$). Nineteen days post-irradiation both the weight and capacity of the dorsolateral prostate to concentrate Zn-65 were well within normal limits. These experiments have illustrated that following whole-body X-irradiation there is a temporarily lowered dorsolateral prostate function to concentrate Zn-65 and a transient arrest in spermatogenesis, denoting the absence of sufficient androgen for these functions.

That this lowered androgen level was due to a pituitary dysfunction is shown in Figure 6. These graphs show that the daily administration of testosterone propionate (100 μg) and chorionic gonadotrophin (1 unit) from 6 to 13 days following X-irradiation prevented the fall in weight and Zn-65 uptake of the dorsolateral prostate noted in untreated X-irradiated rats sacrificed at 13 days post-irradiation. These experiments illustrate that the prostate was able to utilize androgen and that the interstitial cells were able to respond to gona-

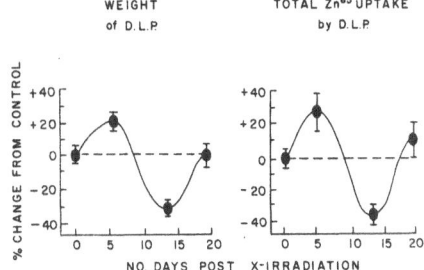

Fɪɢ. 5. Weight and Zn-65 uptake of the dorsolateral prostate of the rat following total-body X-irradiation. Standard errors of the mean are shown (7).

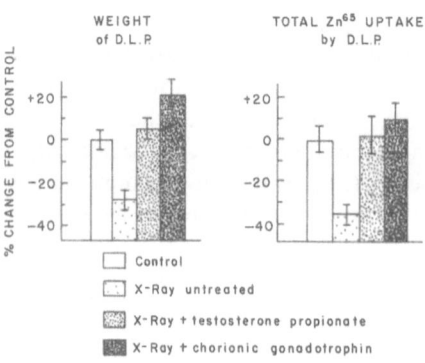

WEIGHT of D.L.P. TOTAL Zn^{65} UPTAKE by D.L.P.

% CHANGE FROM CONTROL

☐ Control
▫ X-Ray untreated
▨ X-Ray + testosterone propionate
▨ X-Ray + chorionic gonadotrophin

FIG. 6. Weight and Zn-65 uptake of the dorsolateral prostate of 13-day post-X-irradiated rat treated with testosterone propionate and chorionic gonadotrophin. Standard errors of the mean are shown (7).

dotrophin. Thus the lowered Zn-65 uptake following X-irradiation exposure was due to a lowered output of gonadotrophin (ICSH or LH) by the pituitary, and not due to any direct damage to the prostate or interstitial cells per se. The ICSH (LH) depletion was only temporary since at 19 days post-irradiation the dorsolateral prostate weight and Zn-65 uptake had returned to normal.

In view of the report that massive doses of ionizing radiation are needed to elicit a change in the pituitary (12) it appears that the lowered ICSH (LH) levels noted in these experiments 13 days following total-body X-irradiation with a dose of only 470 r are not due to any direct effect on the pituitary. Burrows (15) cites various investigations illustrating that castration and damage to the seminiferous tubular epithelium, such as that produced by cryptorchidism, and vasa efferentia ligation, result in characteristic pituitary changes. More recently it has been demonstrated that the derangement of the pituitary elaboration of FSH and LH (ICSH) noted post-castration follows a definite time sequence (16–18). These authors showed that at 10 to 15 days post-castration there was an increase in the pituitary content of FSH with little or no LH (ICSH), whereas 21 days following castration the gonadotrophin content of the pituitary becomes increasingly rich in LH (ICSH). The studies reported here have shown that a minimal discernible effect of X-irradiation, namely, depletion of spermatogonia, is sufficient stimulus to evoke pituitary changes in LH (ICSH) output similar in time sequence to that observed by others following castration. This points up that testicular morphology and function are important regulators of pituitary structure and activity.

With this background in mind we will now proceed to the data gleaned from experiments on microwaves. These experiments were set up to determine if exposure to microwaves of 24,000 mc, 1.25 cm wavelength, energy output of 0.250 w/cm² would produce any disturbance in the pituitary-testes-prostate endocrine chain. The testes of rats were exposed to microwaves a distance of 7.6 cm from the antenna in an environment maintained between 24 and 25°C. Exposures were for 5, 10, or 15 min. At various times following exposure the rats were injected with tracer doses of Zn-65, the animals were sacrificed, and the amount of Zn-65 taken up by the dorsolateral prostate was determined. At the same time the testes were examined grossly and microscopically for damage (8).

Six days following 15-min microwave exposures there were severe third-degree burns of the scrotal skin. The testes showed many opaque areas, hemorrhage, and collapse. Microscopically there was extensive coagulation necrosis of the seminiferous tubules (Fig. 7A). Interstitial and vascular tissues were also involved in the necrosis (Fig. 7B). Twenty-nine days following 15-min exposures the scrotal skin was healing. The testes were fibrotic and reduced in size. Microscopically the tubular outlines were devoid of germinal epithelium (Fig. 7C). The interstitial tissue contained numerous fibroblasts. Very few Leydig cells were present (Fig. 7D).

Six days following 10-min microwave exposure the scrotal skin showed small areas of third-degree burn. The testes were small with a few opaque areas. Microscopically the testes showed focal areas of coagulation necrosis. The majority of tubules showed moderate to severe degeneration with only occasional normal tubules present (Fig. 8A). Of marked interest was the apparent normal appearance of the interstitial tissue (Fig. 8B). Thirteen days following 10-min microwave exposures the scrotal skin burns were healed. The testes were small, with opaque areas present. Microscopically the process of repair was in progress. Tubular debris was not present. Tubules were showing regeneration of cellular elements with active spermatogenesis present (Fig. 8C). Interstitial tissue showed hyperplasia with normal appearing cellular elements (Fig. 8D). Twenty and 29 days following 10-min exposures the histologic picture of the testes was essentially the same as 13 days after exposure.

Six days following 5-min microwave exposure most animals showed no damage to scrotal skin. A few animals showed small

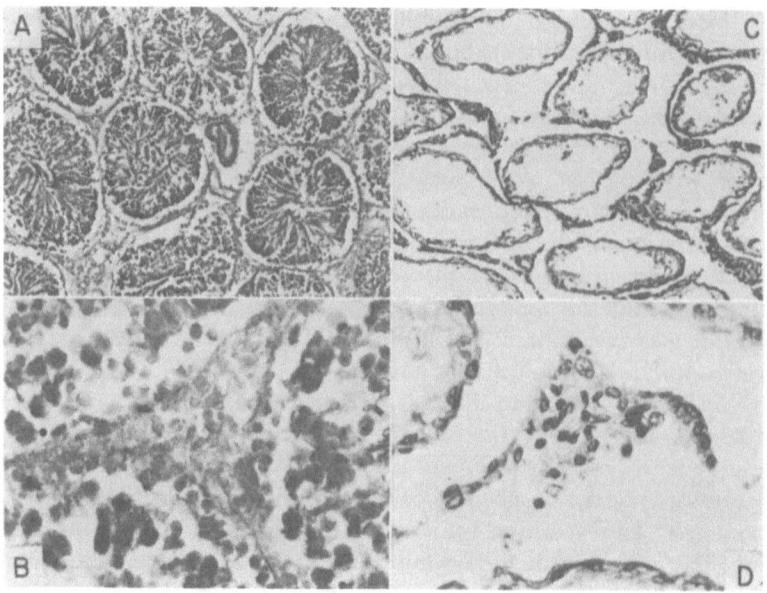

Fig. 7. Testis of rat. (A) Six days following 15-min exposure to microwaves; note the extensive coagulation necrosis of the tubules (H and E, 84X) (8). (B) Six days following 15-min exposure to microwaves; the interstitial and vascular tissues are also involved in the necrotic process (H and E, 450X) (8). (C) Twenty-nine days following 15-min exposure to microwaves; the tubules are completely devoid of germinal epithelium (H and E, 84X) (8). (D) Twenty-nine days following 15-min exposure to microwaves; the interstitial tissue consists mainly of fibroblasts. No mature Leydig cells are evident (H and E, 450X) (8).

second-degree burns. The testes were enlarged, with marked pallor. Microscopically all testes showed moderate to severe edema (Fig. 9A). Most testes showed no tubular damage. A few testes had small areas of tubular degeneration. Interstitial tissue was apparently unaffected (Fig. 9B). Thirteen days following 5-min exposure the scrotal skin was normal and grossly the testes appeared normal. Microscopically the testes showed slight to moderate edema (Fig. 9C). The tubules and interstitial tissue were normal in appearance (Fig. 9D). Twenty-nine days following 5-min exposures the testes were histologically normal.

Figure 10 shows the weight and Zn-65 uptake of the dorsolateral prostate following 5-, 10-, and 15-min exposures to microwaves. Six and 29 days following 15-min exposure to microwaves the capacity of the dorsolateral prostate to concentrate Zn-65 was 60 to 70% lower than control values ($P < 0.001$). Glandular weights were 20% lower than controls ($P<0.01$ for 6 days post-exposure experiment). Six days following 10-min exposure to microwaves the capacity of the dorsolateral prostate to concentrate Zn-65 was diminished 35% ($P < 0.01$). Thirteen, 20, and 29 days after exposure Zn-65 uptake was depressed 55–60% ($P < 0.001$). There was a lowered trend in dorsolateral prostate weight 6, 13, and 20 days following 10-min microwave exposure (not statistically significant). Twenty-nine days after exposure glandular weights were 29% lower than control values ($P < 0.001$). Six days following 5-min exposure to microwaves the capacity of the dorsolateral prostate to take up administered Zn-65 was depressed 30% ($P < 0.01$), and 13 days after exposure Zn-65 uptake was diminished 45% ($P < 0.001$). Twenty-nine days after exposure Zn-65 uptake did not differ significantly from the unexposed

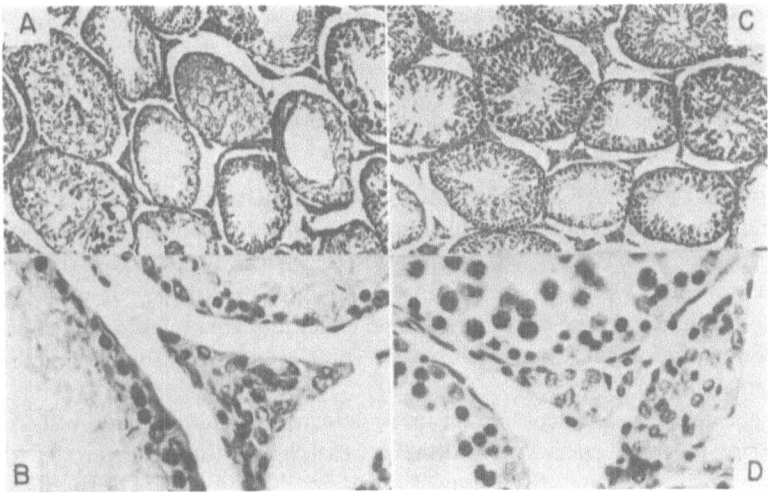

Fig. 8. Testis of rat. (A) Six days following 10-min exposure to microwaves; all tubules show moderate to severe degenerative changes (H and E, 84X) (8). (B) Six days following 10-min exposure to microwaves; note the interstitial tissue is apparently normal (H and E, 450X) (8). (C) Thirteen days following 10-min exposure to microwaves; tubular debris is not present. All tubules show regeneration and active spermatogenesis (H and E, 84X) (8). (D) Thirteen days following 10-min exposure to microwaves; note the hyperplasia of the interstitial tissue with normal cellular elements (H and E, 450X) (8).

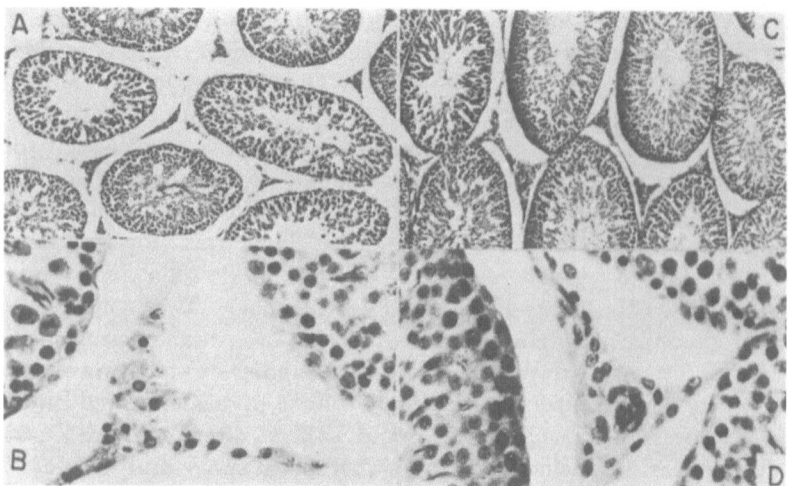

Fig. 9. Testis of rat. (A) Six days following 5-min exposure to microwaves; note the normal-appearing seminiferous tubules and the edema of the intertubular spaces (H and E, 84X) (8). (B) Six days following 5-min exposure to microwaves; note the normal appearance of the interstitial tissue (H and E, 450X) (8). (C) Thirteen days following 5-min exposure to microwaves; the seminiferous tubules are normal with little or no intertubular edema (H and E, 84X) (8). (D) Thirteen days following 5-min exposure to microwaves; the interstitial tissue is normal in appearance (H and E, 450X) (8).

control value. At no time following the 5-min exposure was there any significant alteration in the weight of the dorsolateral prostate. These experiments have indicated that microwaves produce varying degrees of testicular damage, depending upon exposure time. It was also noted that with the 5-min exposures, although there was no evidence of morphological damage, there was interference with the male endocrine system.

Experiments were then set up to determine where in the pituitary-testes-prostate endocrine chain was the defect that interfered with the production or utilization of androgen. The effects of the daily administration of chorionic gonadotrophin (1 unit) and testosterone propionate (200 μg) on the weight and Zn-65 uptake of the dorsolateral prostate following 5- and 10-min exposures to microwaves are shown in Figure 11. In the 5-min exposure studies, which showed Zn-65 uptake depressed 45% from normal ($P < 0.001$), the administration of either gonadotrophin or testosterone restored the Zn-65 uptake to control levels. This experiment indicated that both the prostate and the interstitial tissue of the testes were able to function, and that the cause of the lowered Zn-65 uptake was due to a lack of pituitary gonadotrophin. Dorsolateral prostate weights, though not depressed in the microwave-exposed untreated rats, were significantly increased above normal values by the administration of either hormone ($P < 0.001$). The transient fall in Zn-65 uptake with a later return to normal levels, and the type of response to hormones noted in the 5-min microwave exposure experiments were very similar to the effects noted with the sublethal exposures to X-rays, as demonstrated earlier.

In the 10-min exposure groups, which showed Zn-65 uptake 50% below normal ($P < 0.001$), the administration of testosterone restored

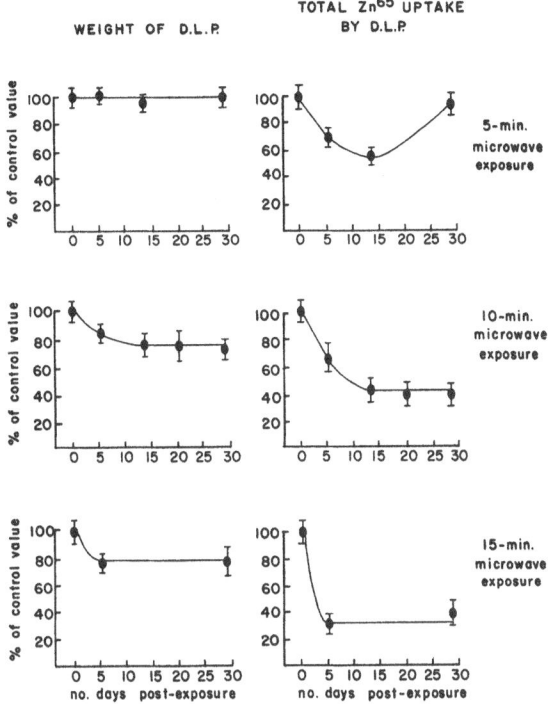

Fig. 10. Weight and Zn-65 uptake of the dorsolateral prostate following 5-, 10-, and 15-min exposures to microwaves. Weight and Zn-65 uptake are expressed as percentage of unexposed control value. Standard errors of the mean are shown (8).

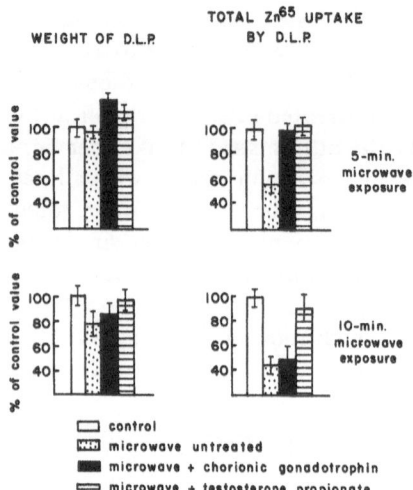

FIG. 11. Effect of administration of chorionic gonadotrophin (1 unit) and testosterone propionate (200 µg) on the weight and Zn-65 uptake expressed as percentage of unexposed, untreated control value. Standard errors of the mean are shown (8).

Zn-65 uptake to control levels, whereas the administration of gonadotrophin was completely ineffective. Dorsolateral prostate weights, depressed in the microwave-exposed untreated rats (though not statistically significant), were restored to normal limits by testosterone, but not by gonadotrophin. These experiments indicated, then, that the prostate is intact, since there was a response to testosterone. However, an apparently normal appearing interstitial tissue is damaged since it is unable to respond to gonadotrophin. To summarize these hormone studies, these experiments indicate that in the 5-min exposed group the cause of diminished androgen output was due to insufficient gonadotrophin output by the pituitary. In the 10-min exposure studies the cause of the diminished androgen output was twofold: (1) a lack of sufficient gonadotrophin output by the pituitary, and (2) a failure of the testicular interstitial tissue to respond fully to whatever trophic hormone was being elaborated.

The next question to be investigated was whether the morphological changes noted in the testes and the endocrine disturbances seen following exposures to microwaves were really due only to thermal effects or due to a combined thermal plus some unknown effect inherent in the microwave range used.

Imig, Thomson, and Hines (1) exposed the scrotal area of rats to 12-cm microwaves, a wavelength which is minimally absorbed by the skin, the brunt of the effect being directly in the testes. In the experiments reported in the present study a wavelength of 1.25 cm was used which is supposedly all absorbed at the skin level (19), the microwaves not penetrating directly to the testes. It is well known that the scrotal skin serves as a thermoregulator protecting the testes from extremes of temperature (20). It is therefore axiomatic that any damage to the scrotum would allow the testis to suffer deleterious effects. In our experiments, using 10- and 15-min exposures in which severe scrotal burns were produced, the protective function of the scrotum was lost, allowing thermal damage to the testis. In these animals morphologic changes in the testis were extensive enough to account for the endocrine interference noted; i.e., diminished weight and reduced Zn-65 uptake of the dorsolateral prostate. In the 5-min exposure studies, which resulted in minor scrotal erythema, no significant testicular damage was evident histologically, but endocrine function was disturbed, as evidenced by depressed Zn-65 uptake by the dorsolateral prostate as long as 13 days after exposure. This brings up the question whether this endocrine impairment could be solely the result of thermal damage to the scrotum and, consequently, to the testis.

Let us briefly consider the results of others concerning the effect of heat derived from various sources on the testes. It has often been reported that heat produces damage to the seminiferous tubules, but that there is no apparent histologic damage to the interstitial tissue (21). Recently work by Steinberger and Dixon (22) emphasized that the production of intratesticular temperatures of 43°C sustained for 15 min resulted in no morphological evidence of testicular damage. Elfving (21) reported experiments in which the scrota of rats were exposed to water bath temperatures of 44.3°C for 20 min which produced intratesticular temperatures of 43.3°C for approximately 15 min. Although 1 week after exposure there was an endocrine hypofunction of the testis, as indicated by a 25% lowered seminal vesicle weight, after 2 weeks glandular weights were normal, and after 3 weeks a slight hypersecretion of androgen ensued.

Figure 12 shows the temperatures that were developed within the testes during 5- and 10-min exposures in our microwave experiments. Intratesticular temperatures were measured with a Thermis-temp Telethermometer. The graph shows that the peak intratesticular temperature developed after 5-min microwave exposure was 41°C, and after 10-min microwave exposure was about 45°C. In our 10-min microwave experiments intratesticular temperatures of the same

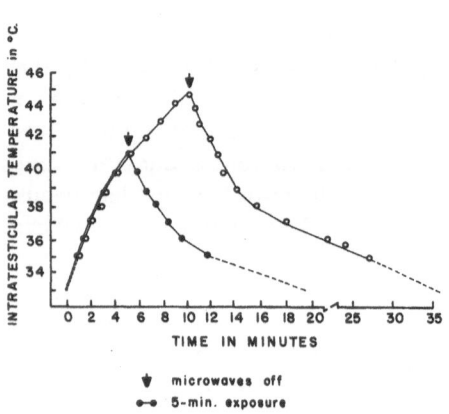

Fig. 12. Mean intratesticular temperatures recorded during and following 5- and 10-min exposures to microwaves (8).

range as Elfving used, 43.0 to 44.5°C, were attained. In our experiments, however, the testes were at this temperature range only for a period of 3 to 4 min, whereas in Elfving's heat experiments the testes were at this temperature range for approximately 15 min. The effects noted on the endocrine system in our microwave experiments appeared to be considerably more severe, however, than the endocrine changes noted in Elfving's heat exposure experiments, i.e., dorsolateral prostate weights were lowered 25–30% and Zn-65 uptake function was diminished 55–60% and showed no return toward normal as long as 4 weeks after exposure. Even in the 5-min exposures to microwaves, in which intratesticular temperatures did not exceed 41°C and remained above 39°C for only 3 to 4 min, there was a 45% diminution in dorsolateral prostate function to concentrate Zn-65 as long as 2 weeks after exposure. Imig, Thomson and Hines (1), working with 12-cm wavelengths, concluded that testicular damage, as evidenced microscopically, resulted from microwaves at a temperature below that necessary to cause injury by infrared exposures, suggesting to these authors that damage may result in part from factors other than heat.

To explore this question further, simultaneous experiments were set up in our laboratory to compare the effects following microwave and infrared exposures. One group of rats was exposed to microwaves for 5 min, at the end of which time the intratesticular temperature developed was 41°C. Another group of rats was exposed to infrared for 5 min. The distance from the infrared lamp to the scrotal area was so adjusted that within 5 min of exposure an identical maximum intratesticular temperature of 41°C was developed. Two-weeks following exposure, both groups were injected with tracer doses of Zn-65, the Zn-65 uptake in the dorsolateral prostate was determined, and the testes were examined grossly and microscopically. The histologic appearance of the testes of microwave and infrared exposed groups was identical. The only histologic alteration noted was a slight edema.

Figure 13 shows the results of the Zn-65 uptake studies in microwave- and infrared-exposed groups. As pointed out before, there was a 45% fall in Zn-65 uptake in the microwave-exposed group ($P <$ 0.001), denoting a break in the pituitary-testes-prostate endocrine chain. In the infrared-exposed group, however, Zn-65 uptake was not altered from control levels, thus indicating that at this temperature range there was no damage to the male endocrine system.

These preliminary experiments indicate that there is a marked difference in the actions of microwaves of 24,000 mc and of infrared on the male endocrine system. Whether or not these experiments suggest an athermal effect of microwaves is not conclusive, but appears suggestive.

This investigation was supported in part by the Air Research and Development Command, Rome Air Development Center, Griffiss Air Force Base, New York. Principal Investigator at the University of Miami, William B. Deichmann, Professor and Chairman, Department of Pharmacology, School of Medicine, Coral Gables, Florida.

The isotope studies in this investigation were supported by a grant from the United States Atomic Energy Commission (Principal Investigator: Samuel A. Gunn, Department of Pathology).

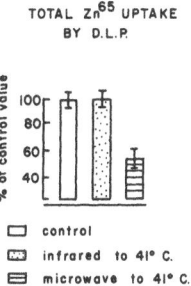

TOTAL Zn^{65} UPTAKE
BY D.L.P.

Fɪɢ. 13. Comparison of the effect of infrared and microwave exposures on the capacity of the rat dorsolateral prostate to concentrate Zn-65. Standard errors of the mean are shown.

References

1. Imig, C. J., Thomson, J. D., and Hines, H. M., "Testicular Degeneration as a Result of Microwave Irradiation," *Proc. Soc. Exptl. Biol. Med.,* 69, 382 (1948).

2. Gunn, S. A., and Gould, T. C., "Hormone Interrelationships Affecting the Selective Uptake of [65]Zn by the Dorso-lateral Prostate of the Hypophysectomized Rat," *J. Endocrinol.,* 16, 18 (1957).

3. Gunn, S. A., Gould, T. C., Ginori, S. S., and Morse, J. G., "Selective Uptake of Zn[65] by Dorsolateral Prostate of Rat," *Proc. Soc. Exptl. Biol. Med.,* 88, 556 (1955).

4. Mawson, C. A., and Fischer, M. I., "Zinc Content of the Genital Organs of the Rat," *Nature,* 167, 859 (1951).

5. Gunn, S. A., and Gould, T. C., "The Relative Importance of Androgen and Estrogen in the Selective Uptake of Zn[65] by the Dorsolateral Prostate of the Rat," *Endocrinology,* 58, 443 (1956).

6. Gunn, S. A., Gould, T. C., and Anderson, W. A. D., "Zn[65] Uptake by Rat Dorsolateral Prostate as Indicator of ICSH Activity," *Proc. Soc. Exptl. Biol. Med.,* 104, 348 (1960).

7. Gunn, S. A., Gould, T. C., and Anderson, W. A. D., "The Effect of X-Irradiation on the Morphology and Function of the Rat Testis," *Am. J. Pathol.,* 37, 203 (1960).

8. Gunn, S. A., Gould, T. C., and Anderson, W. A. D., "The Effect of Microwave Radiation on the Morphology and Function of Rat Testis," *Lab. Invest.,* in press.

9. Bloom, W., *Histopathology of Irradiation from External and Internal Sources.* McGraw-Hill, New York, 1948, p. 550.

10. Shaver, S. L., and Mason, K. E., "Selective Testicular Damage in Rats due to X-Rays," *Anat. Rec.,* 106: 246, 1950.

11. Blair, H. A., *Biological Effects of External Radiation.* McGraw-Hill, New York, 1954, p. 27.

12. Hollaender, A.: *Radiation Biology. Vol. I: High Energy Radiation, Part II,* McGraw-Hill, New York, 1954, p. 1114.

13. Steinberger, E., and Nelson, W. O., "The Effect of Furadoxyl Treatment and X-Irradiation on the Hyaluronidase Concentration of Rat Testes," *Endocrinology,* 60, 105 (1957).

14. Witschi, E., Levine, W. T., and Hill, R. T., "Endocrine Reactions of X-Ray Sterilized Males," *Proc. Soc. Exptl. Biol. Med.,* 29, 1024 (1931-2).

15. Burrows, H., *Biological Actions of Sex Hormones,* 2nd ed., University Press, Cambridge, 1949, p. 111.

16. Hellbaum, A. A., and Greep, R. O., "Qualitative Changes in the Gonado-tropic Complex of the Rat Pituitary following Removal of the Testes," *Am. J. Anat., 67,* 287 (1940).

17. Purves, H. D., and Griesbach, W. E., "Changes in the Gonadotrophs of the Rat Pituitary after Gonadectomy," *Endocrinology, 56,* 374 (1955).

18. Hildebrand, J. E., Rennels, E. G., and Finerty, J. C., "Gonadotrophic Cells of the Rat Hypophysis and Their Relation to Hormone Production," *Z. Zellforsch. rnikroskop. Anat., 46,* 400 (1957).

19. Schwan, H. P., and Li, K., "Hazards due to Total Body Irradiation by Radar," *Proc. I.R.E., 44,* 1572 (1956).

20. Andrews, F. N., "Thermal-regulatory Function of Rat Scrotum. I. Normal Development and Effect of Castration," *Proc. Soc. Exptl. Biol. Med., 45,* 867 (1940).

21. Elfving, G., *Effects of the Local Application of Heat on the Physiology of the Testis,* T. A. Sahalan Kirjapaino Oy., Helsinki, 1950.

22. Steinberger, E., and Dixon, W. J., "Some Observations on the Effect of Heat on the Testicular Germinal Epithelium," *Fertil. and Steril., 10,* 578 (1959).

Effects of Radio-Frequency Energy on Human Gamma Globulin

SVEN A. BACH, ANTHONY J. LUZZIO,
AND ARNOLD S. BROWNELL
Biophysics and Radiobiology Divisions
US Army Medical Research Laboratory
Fort Knox, Kentucky

INTRODUCTION

THE *absorption* of electromagnetic energy in solutions of electrolytes, body fluids, and mammalian tissues has been well-documented throughout the rf and microwave spectrum [Cook (1), Schwan (2), and Cole (3)].

Possible *modes of interaction* of electromagnetic energy with aqueous systems have been treated theoretically [Errera (4)] on the basis of observed anomalous dispersions in dielectric constant and conductivity. Among these are dispersions due to orientation of polar particles in a viscous medium.

Changes have been produced *in colloidal systems* by microwave energy. Van Everdingen (5) in Holland showed in 1946 that starch and glycogen solutions can be altered by 3,000 mc radiation so as to produce complete optical inactivity in both systems and precipitation in the case of starch solutions. These effects could be produced only at certain concentrations or by adjusting the viscosity by adding glycerol. Van Everdingen pointed out the importance of the viscosity term in Debye's equation for relaxation times of polar particles in a viscous medium, reasoning that at a certain combination of viscosity and frequency, maximal energy absorption would occur.

The paper electrophoretic pattern of human gamma globulin alters from a single peak to a double peak when exposed to high doses of X-rays *in vitro*. These electrophoretic changes are accom-

117

panied by changes in antigenic reactivity as measured by titration against the serum of a rabbit immunized against X-irradiated human gamma globulin (unpublished data, Luzzio, USAMRL). This paper describes the changes produced in the paper electrophoretic patterns and in the antigenic reactivity of human gamma globulin when exposed to rf energy in the range from 10 to 200 mc/sec. Twelve hundred and fifty individual exposures have been made, most of them in the range from 10 to 40 mc/sec.

Materials and Methods

All of our exposures were performed in 2.2% solution in normal saline or saline with a phosphate buffer at a pH of 7.6. The solution was placed in a small chamber having two metallic sides, usually silver, which formed the electrodes to which the signal was applied. The electrode dimensions varied from long and narrow (5 × 1 cm) to square (1.5 × 1.5 cm). The most often used chamber had electrodes of 1.5 × 1.6 cm, spaced 3 mm apart. The energy source was a Hewlett-Packard Signal Generator (Fig. 1) controlled by a pulsing

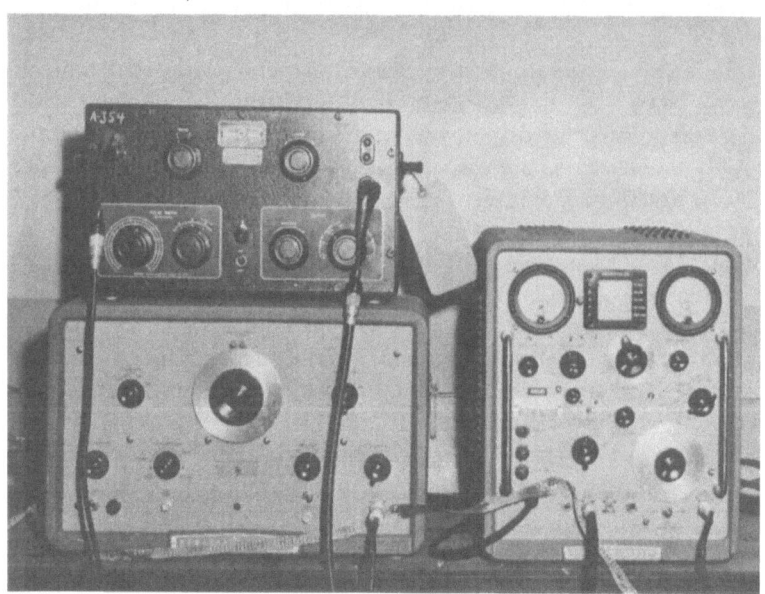

Fig. 1.

Fɪɢ. 2.

system in which pulses could be varied from a few microseconds in width up to 80 μ sec, and the repetition rate from 30 up to 5,000 c/sec. Most of the exposures were of 20-min duration, pulse widths of 10 or 60 μ sec and repetition rates of 500 to 2,000 c/sec.

The pulsed energy from this source was amplified through a cascade of 3 low-power and 2 high-power wide-band distributed amplifiers (Fig. 2). From the final amplifier, the signal passed through a power divider (Fig. 3) which enabled a reading of the power in each direction and hence a measurement of the voltage standing wave ratio. The signal was then passed through a modified π-network which could be tuned for minimum VSWR with a wide variety of loads. The output of this net was placed across the electrodes, one of which was grounded and cooled by a flow of constant temperature water. A shielded IN55A crystal shunted by a 39-kilo-ohm resistor was placed at the point of application of the signal to the ungrounded electrode. The half-wave rectified pulse passed through a transmission line to an oscilloscope where the voltage, pulse width, and repetition frequency could be measured (Fig. 4). The temperature of the solution was continuously recorded from a copper-constantan ther-

Fig. 3.

mocouple held in place with a plastic jig and leading to the recording Brown potentiometer shown. Amplification of the pulse was
excellent in this system. Figure 5 shows the unrectified pulse and
Figure 6 the half-wave envelope resulting from rectification by the
crystal.

Power measurements were made in two ways. In the early experiments using air cooling, the thermal constant of the system was
measured by warming the solution with a cw signal and then recording the change in temperature with time. The equilibrium temperature difference during an exposure and the heat capacity of the
system then give one a measure of the average absorbed power.

With the later water-cooled system, the equilibrium temperature
difference (ΔT) between power on and power off was found to be
proportional to the square of the rectified voltage and to the pulse
repetition frequency (prf).

Extrapolation of the prf vs ΔT curves to zero prf at various voltages gave a measure of the ratio of pulse power to cw power between
pulses and hence the ratio of the pulse voltage to cw voltage. This

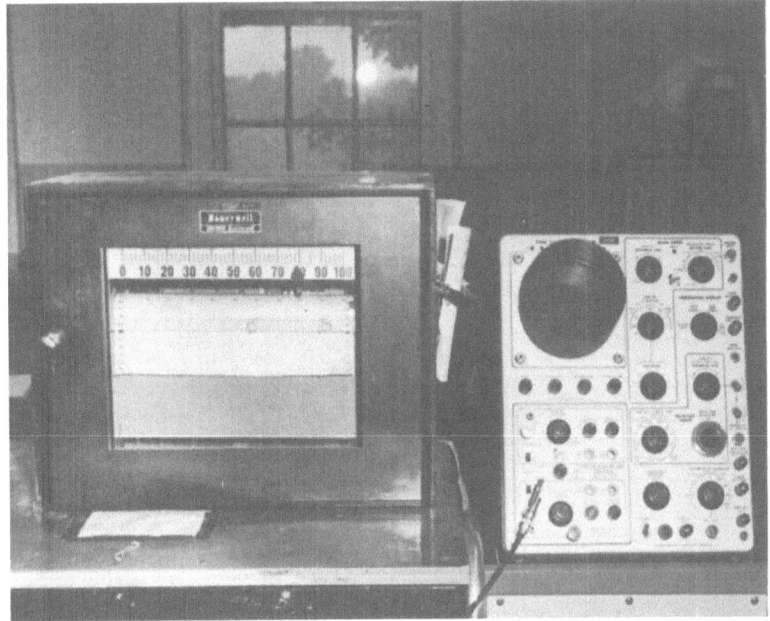

Fɪɢ. 4.

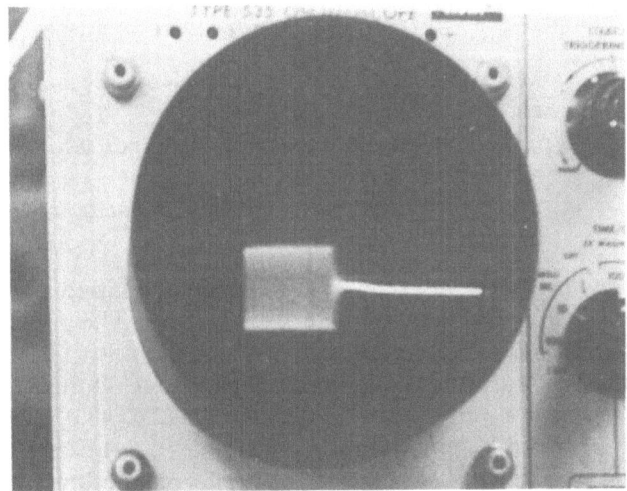

Fɪɢ. 5.

Fig. 6.

was found to be 30 to 1; therefore, except at very low prf's the cw contribution to power was negligible.

A ΔT vs power curve was obtained by measuring the change in temperature of the coolant water at a measured flow rate. Both types of measurement were checked against the calculated power using the following expression:

$$P = (E^2_{eff}/R)fw$$

where E_{eff} = 0.707 times the peak to peak voltage measured on the scope

f = pulse repetition frequency in cycles per second

w = pulse width in seconds

R = resistance of load measured by a capacitance bridge at the operating frequency

These measurements showed, as expected, that this load was essentially a resistive one at these frequencies. The resistivity at 37.5° was found to be 50 ohm-cm.

Gamma globulin solutions were immersed for 30 min in a constant temperature water bath at 33.5, 37.0, 42.0, 48.0, 52.6, 57.2, and 64.2°C. No changes were observed in any but the last, in which a precipitate formed. No precipitate has formed in any of our exposed samples. The electrophoretic pattern of the supernatant solution showed a narrow high peak; no double peaks developed in any of the patterns, the rest of which matched those of the controls precisely. A second experiment with 5, 10, 15, and 20 min immersion at 58, 60, 62, 63, and 64.2°C gave the same results.

Results

The first exposures were at every 10 mc between 10 and 200 mc for 30 min each. At this time water was not used for temperature control. With only air cooling, the temperature rises during exposure were less than 5°C. In Figure 7 the final temperatures attained are seen to be 30°C or less, well below normal body temperature. On paper electrophoresis most of the samples showed no change. However, at 30, 60, 140, 180, and 200 mc, gross changes in the form of a distinct double peak in the electrophoretic pattern were seen. The average powers in this series, based on 5° temperature rises and the thermal time constant of the exposure chamber, were less than 300 mw (60 mw/cm² average power density). There was no correlation between temperature rise and effect.

Since 30 mc seemed a good place to explore in more detail, a number of exposures were made at 1 mc increments in this region. These were all at 13 v rectified, (26V P-P) 1196 c/sec, and 10 μsec pulse width. The field strength was 87 v/cm. Since, by measurement on the capacitance bridge, the load was found to be 3 ohms, the calculated power is then:

$$[(2 \times 13 \times 0.707)^2/3]\,(1196)\,(10^{-5}) = 1.35\,\text{w}$$

or $$1.35/5 = 0.27\ \text{w/cm}^2$$

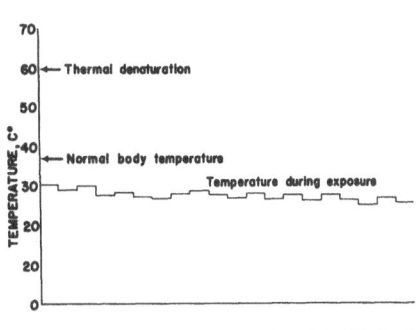

FIG. 7.

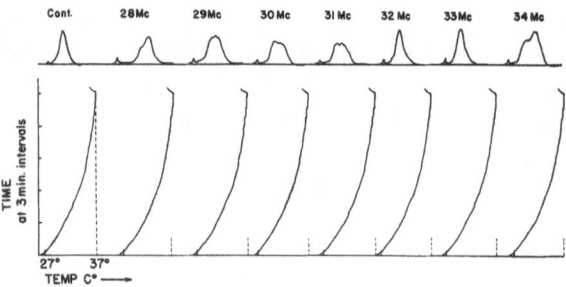

FIG. 8. Effect of frequency shift.

These exposures were all for 20 min. Figure 8 shows the results with the corresponding time-temperature curves. Two things are apparent. One, the electrophoretic changes are not profound, and two, the band width under these conditions looks like 4 or 5 mc out of about 30 or about 13–16% of the frequency.

A repetition gave positive results at 29, 31, and 34 mc, the changes being greater. A limited range at about 5% frequency increments was then explored. The temperature was not controlled and in all these exposures varied between 30 and 40°C.

Each of the exposures within the individual runs was at the same voltage, pulse width, pulse repetition rate, power, and duration, only the frequency being changed. Exposures were randomized within each run.

Some typical exposure parameters for these runs are as follows:

w/cm², average power density	Pulse repetition frequency	Pulse width, μsec	Field strength P-P v/cm	Exposure time, min
3.4	2000	10	240	20
4	3300	10	200	20
3.3	1780	70	93	20
2	1667	10	200	20
4.8	1000	10	400	20

It became obvious from the results that we were not controlling conditions well enough between exposures and that 5% frequency increments were too large, since sometimes we missed obtaining an effect. After making six sweeps at 5% increment between 10 and 41 mc it was, however, apparent that the successful exposures were grouped in some sort of pattern.   At about this time we ran across Van Everdingen's experiment with starch and glycogen and decided to use Debye's equation as a guide to possible frequency shift with temperature.   If one takes the empirical equation for viscosity of a liquid

$$\ln \eta = (a/T) + b$$

and combines it with Debye's equation

$$\tau = 3v\eta/kT$$

for a spherical particle then

$$f = cT \exp\left[(a/T) + b\right]$$

On differentiation, the change in frequency with temperature in a water solution can be calculated.   This turns out to be 2.42%/°C. We now went back to the data from an earlier experiment at 22°C in which we had cooled both electrodes with diesel fuel.   This kept the temperature constant with a very low gradient.   In this experiment we had obtained effects at 10, 15, 20, and 25 mc, which seemed to indicate that harmonics of 5 mc were involved.

If one starts with 5 mc at 22°C and then calculates the harmonics of the resulting frequencies at 30 to 40°C they appear as shown in Figure 9.   If one then assumes a certain probability of success at an effective frequency under the exposure conditions for this series, say 0.25 (which was the experience figure), and further assumes that the points of overlap in the harmonics will increase your chances of success ($P_2 = 1 - q^2$), then the theoretical probabilities of success appear as shown at the bottom of Figure 9.

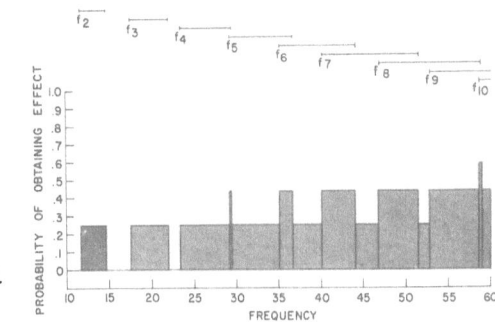

FIG. 9.

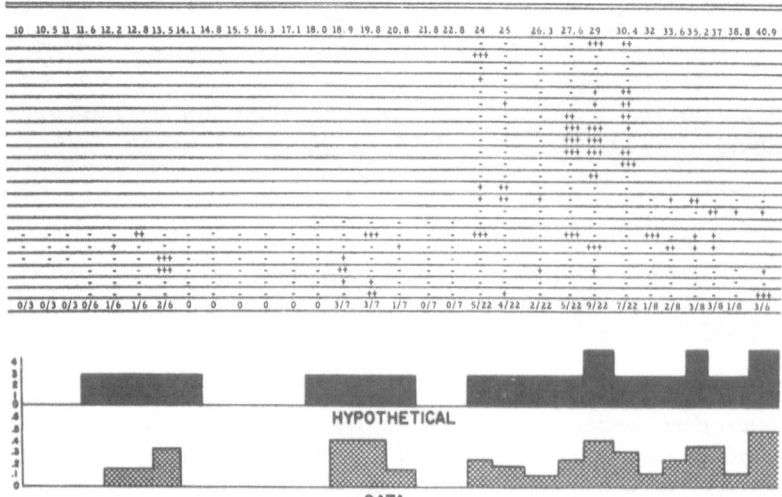

Fig. 10.   Frequency of positive results at 5% increments of frequency 10–41 mc.

Taking these probabilities and applying them to the actual frequencies of exposure (Fig. 10), one obtains a theoretical pattern of successes marked "Hypothetical" in this figure.

The summary of the data is shown on the bottom of Figure 10. Since the total number of exposures was small in part of the spectrum, the data were then grouped in 10% bands (Fig. 11). Here a rather striking agreement is seen between theoretical and actual results. These considerations would then lead one to surmise that harmonics of around 6 mc were involved in this process, the exact frequency depending on temperature. Since our equipment did not go below 10 mc, we went to the f_2 harmonics (calculated at between 12.4 and

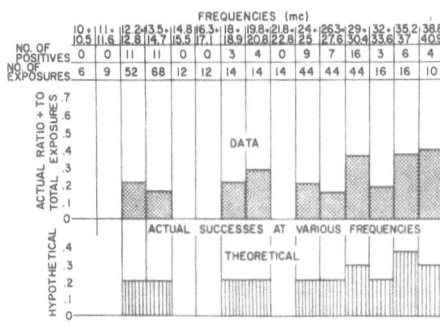

Fig. 11.

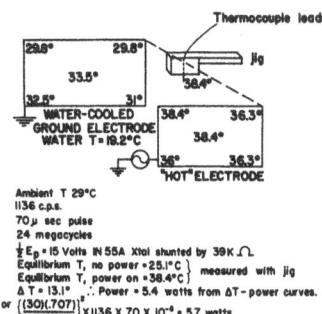

Ambient T 29°C
1136 c.p.s.
70 μ sec pulse
24 megacycles

‡ E_p = 15 Volts IN 55A Xtal shunted by 39K Ω.
Equilibrium T, no power = 25.1°C ⎫ measured with jig
Equilibrium T, power on = 38.4°C ⎭
ΔT = 13.1° ∴ Power = 5.4 watts from ΔT - power curves.
or $\{\frac{(30)(.707)}{6.25}\}$ X1136 X 70 X 10⁻⁶ = 5.7 watts

Fig. 12.   Temperature gradients in a water-cooled cell.

13.5 mc) and exposed in 100 kc increments, once at a relatively high power and high temperature gradient in the cell (Fig. 12), and again at about one half the temperature gradient. These runs (Fig. 13) showed the expected result, i.e., a narrowing of the band width without decreasing the effect where it did appear (this might be because even at lower power and fewer pulses per second, more molecules would be at the right frequency and temperature combination during the exposure).

However, a few more series showed that though we were improving our successes, we were still missing something (Fig. 14).

A review of the exposure data showed that some of the exposures were made during warm wet days and others during cool dry days. We obtained relative humidity readings for all daylight hours during the period of this experiment and grouped the data according to relative humidity during each individual exposure with the results shown in Figure 15. It is seen that on the dry cool days the frequency pattern is to the left, and on warm wet days it seems to shift to the right. The possible mechanism, based on Debye's equation, is shown in Figure 16.

FREQUENCY Mc/sec

	12.4	12.5	12.6	12.7	12.8	12.9	13.0	13.1	13.2	13.3	13.4	13.5
High temperature gradient. 60 μsec 1040 c.p.s.	+++	+++	+++	++++	+++	-	-	-	-	-	-	-
				6 - 7°C								
Low temperature gradient 60 μsec 525 c.p.s.	-	++++	++++	+	-	-	-	-	-	-	+	-
				3 - 3.5°C								

Voltage (rectified)
~ 20 Volts for
both runs

Fig. 13.   Effect of lowering the temperature gradient in a cell.

Bach et al.

FREQUENCY MC/SEC

Run #	12.4	12.5	12.6	12.7	12.8	12.9	13.0	13.1	13.2	13.3	13.4	13.5
High power (1)	+	-	-	-	-	+++	-	-	-	-	-	-
5.2 watts (2)	-	-	-	-	-	+++	-	+	++	-	++	+++
(3)	+++	+++	+++	++++	+++	-	-	-	-	-	-	-
(4)	-	++++	++++	+	-	-	-	-	-	-	+	-
Lower power (5)	-	-	-	-	-	-	-	-	-	-	-	++++
3.1 watts (6)	-	-	-	-	-	-	-	-	-	-	-	-
(7)	-	-	-	-	-	-	-	++	-	-	-	-
(8)	-	-	++	-	-	-	-	-	-	-	-	-
(9)	-	-	-	-	-	-	-	+	-	-	-	-

	2/9	2/9	3/9	2/9	1/9	2/9	0/9	3/9	1/9	0/9	2/9	2/9

Fig. 14. Results at 100 kc increments.

We then placed the cell in a humidified incubator where the air temperature was maintained at 31°C. This provided more uniform average temperatures in the cell, and cut down evaporation.

It was calculated from the data at 35.1° that at 37.5° one should obtain effects at 13.1, 13.2, 13.3, 13.9, 14.3, and 14.4 mc.

Our next series at 37.5° at 100 kc increments between 12.8 and 14.5 mc showed positives at 13.1, 13.2, 13.3, and 14.3 mc as predicted. After several replications it appeared that we were getting a frequency shift of about 100 kc and that the band width under these conditions was narrower than 100 kc.

The frequency increment was then cut to 20 kc and with careful crystal calibration of the generator between each exposure, we explored the range between 13.00 and 13.34 mc. We obtained beautiful double peaks at 13.12 and 13.32 mc. Six replications showed a

FREQUENCY Mc/sec

Rel. humidity:	12.4	12.5	12.6	12.7	12.8	12.9	13.0	13.1	13.2	13.3	13.4	13.5
< 54	+	-	-	-	-	-	?	-	-	-	-	-
55 - 60	+++	+++	+++	?	-	-	-	++	-	-	-	-
61 - 65	-	++++	++++	++++	+++	+++	-	-	-	-	-	
66 - 70	-	-	-	0	-	0	-	-	-	+	-	
>71	-	-	-	-	-	++++	-	+	++	-	++	++++ / ++++

O = No data

Fig. 15. Relationship of frequency to relative humidity.

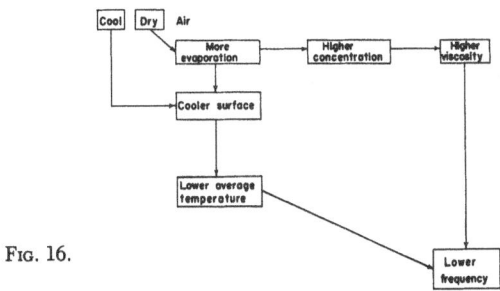

FIG. 16.

similar pattern with 20–40 kc shifts in the effective frequencies about the points 13.10, 13.20, and 13.30 mc.

The actual electrophoretic patterns and the associated time temperature curves for the first run are shown in Figure 17. It is immediately apparent why this effect is so exasperatingly elusive. At 20 kc on either side of this gross change, there is no hint of anything occurring except perhaps the high peaks. This is a frequency increment of about 0.15%, corresponding to a temperature shift of 0.06°C.

Now this electrophoretic change indicates a profound molecular change. A limited number of dialyses in fact show that a group of molecular weight less than 8,000, containing tyrosine, is broken off when the double peak appears in the electrophoretic pattern.

A more sensitive indicator of change was found in the form of rabbit serum containing a precipitin for human gamma globulin. A serial 2-fold dilution technique was used.

To our surprise significant *increases* in titer occurred in samples exposed at 13.2, 13.3, 13.5, and 14.4 mc whereas the others at every

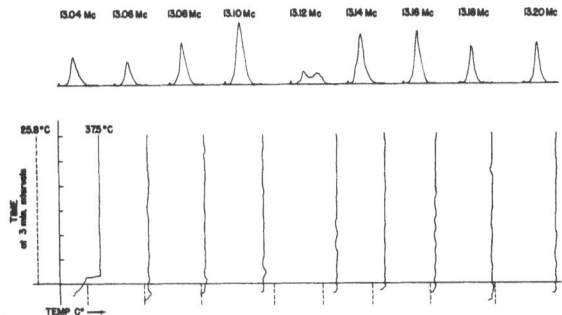

FIG. 17. Narrow banding effect of constant temperature exposure.

100 kc, between 12.8 and 14.5 mc, showed no significant differences from the controls.

These high titers occurred on each side of the frequency at which the electrophoretic double peak appeared. In the sample showing a double peak (Fig. 17) the titer was not significantly different from the control, while at 60 kc below and 120 kc above, the titers were sixteen times that of the controls.

Figure 18 shows the results of two runs at 100 kc increments, at 37.5°C, between 12.8 and 14.5 mc.

Taking a titer of 4 times the control as significant, it appears that there are 3 regions in this part of the spectrum centering at 13.2, 13.5, or 13.6, and 14.4 mc where the changes occur. These agree with the regions of electrophoretic change at this temperature, though as already mentioned, the individual samples showing double peaks do not have a high titer, but rather are adjacent in frequency to the high titer samples.

The latest experiment, still in progress, was to determine the minimum field strength and/or power for detectable effect.

A run was made at every 10 kc to pick an appropriate region between 13.04 and 13.39 mc for the reduced power experiment. Figure 19 shows the results of the 10 kc run.

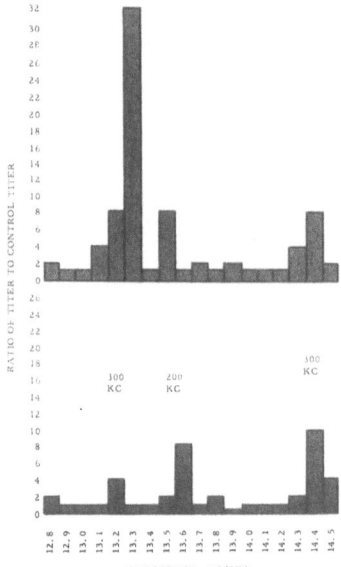

FIG. 18.

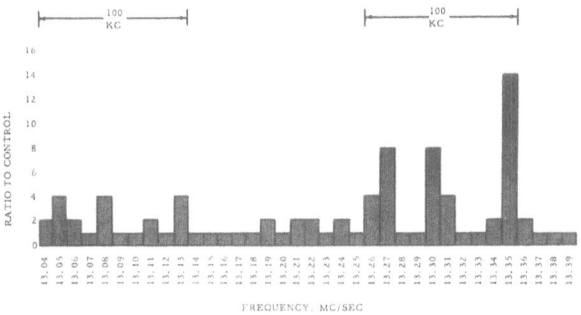

FIG. 19.

Figure 20 shows the fragmentary data at reduced power. It is seen that at 1/150 of the power used for the 10 kc series, an effect on the antigenic reactivity is still produced. This is at an average power density of 13.4 mw/cm². Exposure time was 1 hr. No detectable temperature rise occurred.

It appears from our results that one can alter the electrophoretic pattern and the antigenic reactivity of human gamma globulin by exposing it *in vitro* to rf energy of the proper frequency. The frequency depends on the temperature of the solution.

The temperature dependence of frequency appears to be of the order predicted by Debye's equation for relaxation times of polar particles in a viscous medium, 2.4%/°C for water in the temperature range 30–40°C.

Mass heating of the medium has no relationship to the changes. Neither does average power absorbed. The changes can be produced under widely varying conditions of voltage, power, pulse width, and pulse repetition rate, provided the frequency is suitable. At a non-effective frequency, very high powers and field strengths (4.8 w/cm² and 400 v/cm) will not produce any changes whereas at an effective

	NO EFFECT AT WRONG FREQUENCY	POSITIVE EFFECT AT RIGHT FREQUENCY	NO EFFECT AT RIGHT FREQUENCY
Field strength, volts/cm	400	87	43
Pulse width, microseconds	10	10	60
Pulse repetition frequency, c.p.s.	1000	1196	1667
Average power density, watts/cm²	4.8	0.27	0.73
Peak instantaneous power density, watts/cm²	480	22.6	7.3

FIG. 20.

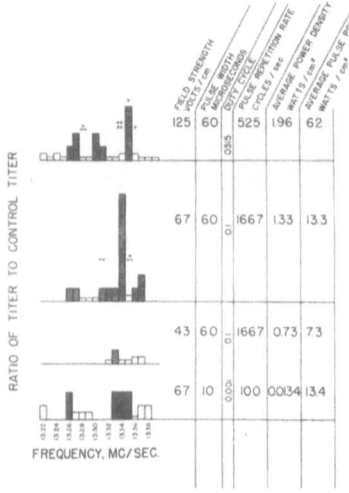

FIG. 21. Effect of reducing power.

frequency, even a very low average power level (13.4 mw/cm²) is sufficient to produce changes.

From the changes in antigenic reactivity one might theorize that this energy produces some degree of unfolding of the protein helix, laying bare more specific combining sites.   With increased exposure to an effective rf frequency, the molecules (or some of them) may become so grossly altered as to lose their specificity.

At 37.5°C in normal saline with phosphate buffer at a pH of 7.6, in the portion of the spectrum studied, the effective frequencies for human gamma globulin are 13.1, 13.2, 13.3, 13.5, 13.6, and 14.4 mc. These may be the second harmonics in a series of harmonics which are also effective.

References

1. Cook, H. F., "A Comparison of the Dielectric Behaviour of Pure Water and Human Blood at Microwave Frequencies," *Brit. J. Appl. Phys., 3,* 249 (August 1952).

2. Schwan, H. P., and Li, K., "Hazards Due to Total Body Irradiation by Radar," *Proc. I.R.E., 44,* 1572 (November 1956).

3. Cole, K. S., and Cole, R. H., "Dispersion and Absorption in Dielectrics. I. Alternating Current Characteristics," *J. Chem. Phys., 9,* 341 (April 1941).

4. Errera, J., "Les milieux colloïdaux et les ondes hertziennes de haute fréquence," *Acta Union intern. contre cancer, Paris, 4,* 195 (1939).

5. Van Everdingen, W. A. G., "Sur l'alteration moléculaire et structurale par irradiation avec des ondes hertziennes de 16 et 10 centimetres (1875 et 3000 MHZ) transformations moléculaires; influence sur le cancer du goudron; metabolisme hépatique et probléme du cancer," *Rev. belge sci. méd., 17,* 261 (October 1946).

Longevity and Cellular Studies
With Microwaves

SUSAN PRAUSNITZ, CHARLES SÜSSKIND,
AND PAUL O. VOGELHUT
University of California
Berkeley, California

ONE OF OUR STUDIES during the last year has been the investigation of the effect of chronic irradiation upon the physiology and longevity of the mouse.

A colony of 300 male NAMRU albino mice was kept in a cabinet, the temperature of which was thermostatically controlled between 21 and 24°C. Each cage containing 10 mice constituted a unit. Each unit was irradiated in a polystyrene cage which was suspended above the radiating horn and rotated at the rate of 1 rpm to minimize the effect of multiple reflections. Two hundred mice were irradiated 5 days a week for 4.5 min at a power density of 0.109 w/cm^2. The average body temperature rise for this dose is 3.3°C. One hundred control mice received the same treatment except that the power was not turned on. The longevity irradiations began on 1 October 1959. To date there have been 17 deaths (8.5%) in the irradiated group and 10 deaths (10%) in the control group.

When an animal died, it was autopsied and key tissues were prepared for extensive histological examination. The tissues studied included liver, spleen, thymus, lymph nodes, kidneys, adrenals, gut, lung, and testes. A lymphoid leucosis was evident in a number of cases where the lymphatic organs were found to be enlarged, but this was true of both irradiated and control mice. A white granular material filling the peritoneum and adhering to the liver was found in 6 mice at different times, but again, this finding was evident in both groups. It is suspected that this material, which was accompanied by high white counts, is a liver abscess. Microscopic examination revealed abscesses in kidneys and liver in both groups. In some cases the lungs were congested and indicated that the mouse had died of bronchial pneumonia.

Though the causes of death were varied, no criterion for separating the irradiated from the control mice has so far been discerned. It also appears that over a period of 33 weeks of irradiation essentially no difference between the weight changes of the irradiated mice and the controls occurred.

Five per cent of both groups were sacrificed during the 8th month of irradiation in order to study the effects of the radiation on the condition of a sampling of the surviving mice. Tail blood was taken from 10 irradiated and 5 control animals for total and differential white blood counts. The mice were then given a strong dose of ether and autopsied. Tissues removed for histological examination were the adrenals, kidneys, testes, brain, liver, duodenum, and spleen. A lung tumor was found in one irradiated mouse; another irradiated animal showed bone tissue in the spleen. This particular animal had had an infection which subsequently healed, as revealed by a temporary elevated white count.

On the basis of six conditions evident in a large proportion of the mice, criteria were set up for separating the controls from the irradiated. These criteria are:

1. Lymphoid infiltration in the brain, liver, kidney, and/or duodenum.

2. Seroid deposits in the interstitial cells of the testes.

3. Presence of anomalous basophilic cells in the seminiferous tubules of the testes.

4. Absence of vacuoles in the adrenal cortex.

5. Presence of congestion in the kidney.

6. Results of total and differential white blood counts.

None of the above categories produced a correlation between the condition and either irradiated or control group. The only task remaining is the examination of bone marrow smears made at the time of the autopsies.

In view of the histological findings of the dead mice and those in the sacrifice series, it appears so far that 8 mo of exposure to a power density of 0.109 w/cm^2 produces no deleterious effects on the mice. This experiment will continue under the same conditions for at least 12 mo. Another sacrifice series will be made in 6 mo.

The second type of investigation that was undertaken at the University of California at Berkeley was a study of the possible use of microwaves as a research tool in biology.

In a number of biological phenomena and in some crystallization and polymerization effects, there is evidence of an attractive intermolecular force which is specific in that identical or nearly identical molecules interact more strongly than non-identical ones. Many investigators have pointed out that biological evidence indicates the existence of such an attraction acting over distances at which the interacting molecules are not in contact, and they have urged physicists to investigate whether any of the known intermolecular forces was capable of accounting for such a phenomenon. Jehle investigated the conditions under which the London van der Waals' dispersion force between particles immersed in a medium will constitute such a specific attraction.

Observed specificity effects may involve a variety of forces. If the interacting molecules can approach one another closely, the most important interactions are bond and bridge formation, in particular those between complementary structures, electrostatic interaction of complementary charge distributions, and van der Waals' stabilization of those complementary structures which permit a closest fit. These interactions account for many biological specificity effects and usually play a decisive role in crystal formation. The simple lock and key picture characterizes these interactions.

On the other hand, if the interacting molecules are not in direct contact, but are still fairly close, the London force might constitute an important specific interaction. In this case identical structures, rather than complementary ones, would tend to aggregate. The London interaction would be sharply specific in the case of macromolecules whose representative oscillators had sufficiently large polarizabilities and frequencies covering a wide range.

Numerous investigators have observed a specific effect of electromagnetic radiation on tissue cultures, microorganisms, enzyme systems, and protein molecules. A theoretical understanding of the effect produced by such radiation has not been achieved as yet. To gain insight into the interaction mechanism between electromagnetic radiation and complex biological systems, the best approach seems to be an attack on the molecular processes involved.

The proteins in a living organism are surrounded by an aqueous atmosphere; all biologically important functions of these macro-

molecules are based on this fact.   Since observed effects of electromagnetic radiation appear to occur at the macromolecular level
(specific action on protein molecules) as well as on the unicellular
level, the conclusion is reached that the effects are mediated by the
protein molecules in their native state, suspended in an aqueous
phase.

A molecule interacts with its neighbor molecules by virtue of the
electron and proton charge distributions on the molecule.   Charge
fluctuation forces arise when the fluctuations of charge distribution
of a molecule are brought about by quantum-mechanical motion, or
by temperature motion, of energy kT.   These charge fluctuations in
a molecule set up energetically favorable charge distributions in the
adjacent molecules because the molecules are electrically polarizable.
The interaction of the fluctuating charge distribution in one molecule with the induced charge distribution in the other, vice versa,
leads to attraction.

If one is concerned about electronic oscillators and their quantum-
mechanical or thermal charge fluctuations, then the forces are of the
London-Eisenschitz-Wang type (1).   If Brownian motion of the
protons on the surface of the molecules is the relevant phenomenon,
the forces are of the Kirkwood-Shumaker type (2).

As long as the ground states of the electronic oscillators are all
that have to be considered (that is, no excited states are in thermal
reach), the calculation is the simple London calculation: the oscillators in the two molecules are coupled by electric dipole-dipole interactions that bring about normal-mode frequency displacements and
corresponding level shifts.   The normal-mode frequencies each
contribute a zero-point energy.

If excited oscillator states have to be considered, one may have
the situation of one excited molecule approaching another unexcited
molecule (3).   Resonance brings about interaction energies proportional to $\pm R$ inverse cube of separation.   If one considers molecule
pairs embedded in a liquid, this medium represents a temperature
bath.   The statistical averaging over excited states is provided for by
a Gibbs ensemble of such molecule pairs distributed over excited
states according to a Boltzmann distribution.   This averaging results
in an interaction free energy per oscillator pair in the classical limit
of low-frequency oscillators (classical formula) (4).

The actual situation implies participation of both classical and
quantum frequency ranges, and thermally possible states have to be

considered. The interaction energy is again proportional to the inverse sixth power of separation. In this general case, the molecules are representable by many oscillators, which have their natural frequencies distributed all over the classical and quantum regions and whose spatial orientations show anisotropies. The calculation may be accomplished by an evaluation of the separation-dependent normal-mode frequencies of molecule pairs, and of the Boltzmann distribution over the quantum states of those harmonic normal-mode oscillators of a dissimilar molecule pair.

When two identical molecules immersed in a liquid medium want to get together as nearest neighbors, some medium molecules have to give way. Thus it is a differential effect between the interaction energy of two identical molecule pairs and non-identical molecule pairs that has to be considered. This effect amounts to a rearrangement energy, well known in the theory of mixtures; the entropy of mixing would have to be considered only when arrangements of a large number of molecules of dissimilar types are to be studied.

The object of the present experiment is to measure the above-mentioned rearrangement energy by measuring the number of bound water molecules that are displaced during the process of rearrangement and computing the amount of energy from the number of bonds that have to be broken in order to accomplish this rearrangement.

Previous descriptions of the structure of water and the effect of solutes on it have been essentially qualitative. Haggis (5) and others have attempted to formulate a quantitative statistical approach. It included a discussion of the bound water of hydration around polar molecules and ions in the interpretation of dielectric-dispersion measurements.

The frequency variation of the dielectric constant of an aqueous solution of polar molecules shows two dispersion regions, one due to the relaxation of the solute and one due to that of the water molecules. Microwave measurements in the wavelength region 1–10 cm cover the range of the water relaxation. Provided that the two dispersion regions are widely separated or that the solute makes only a negligible contribution to the dielectric constant in the microwave region, it is possible from microwave measurements to obtain information about the following:

1. The dielectric constant of water between the two relaxation regions. This constant can be interpreted in terms of the water ir-

rotationally bound to the solute, that is, not free to rotate in the electric field.

2. The time of relaxation of the water molecules, which one can directly relate to the water structure in the solution.

The low-frequency dielectric constant of the water present in the solution may be used to determine the extent to which water molecules around an ion or organic group are prevented by local fields from becoming oriented in an external field. Since these water molecules are restricted only insofar as the local field is stronger than the field due to the neighboring water molecules, the dielectric effect should be closely related to hydration. Considering the water structure in detail, one arrives at an effective number of water molecules irrotationally bound to a solute molecule. The estimate of this number from the static dielectric constant must depend on a dielectric theory of mixtures. The problem of a mixture of polar liquids in the region between their relaxations is equivalent to that of a mixture of polar and nonpolar liquids. At microwave frequencies the protein with its bound water may be considered as being a spheroidal cavity of low dielectric constant in the water, the error being small. The dielectric constant ϵ_s is obtained as a function of the imaginary and real part of the dielectric constant at microwave frequencies.

The following outlines the method of measurement of the two parts of the dielectric constant. Since the rearrangement in the enzyme-substrate solution takes place rather rapidly, a special double-sweep technique is used for the analysis of time-dependent cavity characteristics in conjunction with a rapid flow and mixing system.

A cavity resonator method for the measurement of the dielectric constant and loss of small samples has been developed by Sproull and Lindner (6). This method is characterized by convenient measurement techniques, simple calculations, and direct extension to the measurement of the shunt impedance of a cavity.

Slater (7) has developed a perturbation theory that gives the change in resonance frequency and loaded Q of a cavity due to some small change in the cavity. His equations are useful if this perturbed electric field can easily be obtained from an unperturbed field. As an example, he discusses the case where the surface of the dielectric is everywhere parallel to the electric field. With the further restriction that the dielectric ends with the field lines on the walls of the cavity, the two electric fields are assumed to be equal.

In the present arrangement a small cylindrical sample is placed parallel to the electric field by being inserted through the holes in the wider walls of a rectangular cavity operating in the TE_{106} mode, and from suitable equations the real and imaginary constants of the dielectric constant can be obtained.

In the present scheme for reflection measurements one utilizes both a swept probing signal and a swept receiver to give a resulting pulse whose shape is sensitive to cavity resonance. Frequency shift is determined from the shape of the pulse and its position along the received signal trace.

From the shape of this curve the Q of the cavity is computed. The need for tuning oscillators is virtually eliminated and the minimum point of resonance is sharply defined; moreover, the method is applicable to situations where intermediate frequency is such that video generators would not have sufficient frequency discrimination. Frequency-shift detection is of the order of 10 kc in the 10-kmc operational range. The enzyme-substrate solution injection into the cylindrical sample holder in the cavity (see below) is triggered at zero time at a signal from the master generator. At some time T later the cathode-ray oscilloscope sweep is triggered to give visual display and to sweep the local oscillator synchronously. The amplitude of the modulation signal to the local oscillator, and hence the dispersion of the receiver response, is adjusted by a divider across the cathode-ray-oscilloscope sweep output. At some phase t later, a third triggering pulse from the master generator is sent to a sawtooth generator, set for single-shot operation, which applies a sweep signal to the search oscillator.

The rapid-flow apparatus is designed along the lines suggested by Chance and Legallais (8). They describe a regenerative-flow apparatus that is capable of continuous and intermittent operation. By this technique one is able to detect intermediates 10–30 msec after their formation. In principle the apparatus consists of motor-driven syringes, a mixing chamber, and the observation tube that constitutes the perturbation in the microwave cavity.

The above is an outline of a new instrument that uses microwaves as an analytical tool in the investigation of biophysical problems associated with enzyme reactions. The use of this instrument may be extended to studies of specificity of macromolecular arrays by subjecting the solution under investigation to electromagnetic radiation of diverse wavelengths and thus introducing noise into the reaction system.

References

1. Yos, J. M., Bade, W. L., and Jehle, H., "Specificity of the London-Eisen-schitz-Wang Force," *Proc. Natl. Acad. Sci.*, *43*, 341 (1957).

2. Kirkwood, J. G., and Shumaker, J. B., "Forces between Molecules in Solution Arising from Fluctuations in Proton Charge and Configuration," *Proc. Natl. Acad. Sci.*, *38*, 863 (1952).

3. Mullikan, R. S., "Resonance Interaction of Normal and Excited Atoms," *Bull. Am. Phys. Soc.*, *4*, 173 (1959).

4. Yos, J. M., Bade, W. L., and Jehle, H., Ibid.

5. Haggis, G. H., Hasted, J. B., and Buchanan, T. J., "The Dielectric Properties of Water in Solutions," *J. Chem. Phys.*, *20*, 1452 (1952).

6. Sproull, R. L., and Lindner, E. G., "Resonant Cavity Measurements," *Proc. I.R.E.*, *34*, 305 (1946).

7. Slater, J. C., "Microwave Electronics," *Rev. Mod. Phys.*, *18*, 441 (1946).

8. Chance, B., and Legallais, V., "Regeneration and Circulation of Reactants in the Rapid-Flow Apparatus; Pt. 2: Practical Designs," *Discussions Faraday Soc.*, *17*, 123 (1954).

Phantom Experiments with Microwaves at the University of Rochester*

The University of Rochester School of Medicine and Dentistry
Department of Radiation Biology
Rochester, New York

AT AN EARLIER MEETING of microwave investigators, a number of experiments with phantoms simulating different species of animals were presented. These experiments were designed to demonstrate the time and temperature relations which must exist between small, medium, and large animals, when these are exposed to microwave power at different frequencies. These studies indicated that a small volume phantom, comparable in size with a rat, requires a shorter time to be heated to a given temperature than a relatively large phantom, comparable to a dog, when both are exposed to the same power flux from a microwave generator. Several valid reasons for such behavior are obvious.

First, from considerations of thermodynamic principles, the heat capacity of the smaller volume phantom requires less caloric heat input than the larger volume phantom for the same degree of heating.

Secondly, the degree of penetration of microwaves, although the same in both small and large phantoms or animal species, involves in the small phantom a greater total percentage of volume than in the larger phantom where only a quasi-superficial outer layer is profused with microwave power. This phenomenon is explained by the fact that, for example, 3000 mc microwaves penetrate either tissue or water to a depth of about 2.5 cm. A rat has approximately a diam-

* Based on work done for the United States Air Force under Contract #AF30-(602)-1813 dated 3-1-58.

143

eter of 5 to 6 cm, and therefore, its total volume represents fairly complete absorption.   Conversely, a dog may be 20 to 25 cm thick in some body sections, and thus a secondary process of indirect heating by the blood flow from the superficial vascular bed into the interior depth is responsible for the slower elevation of temperature. From this one may infer that if a man were exposed to the same power density, his rate of temperature rise would be slower than either rat or dog simply on the basis of larger volume.

The third factor which controls the rate of heat induction has already been discussed as the parameter of frequency and its effect on the depth of penetration.

From these earlier experiments, it would appear that if only thermal effects were observed, the hazard to man from exposure to microwaves should be less at higher frequency than at the lower frequencies, and therefore, the single numerical value of 10 mw/cm^2, which is presently uniformly accepted as the maximum permissible exposure, might require some modification with regards to the frequency of the microwave power.   It can also be pointed out that further modification may be in order on the basis of considering the time element during which 10 mw/cm^2 is regarded as safely applicable to humans.   As is well known, no finite answers to these questions have been attained.

Some other studies in progress can be reported at this time.   Our biological investigators have often posed the question concerning the vertical and horizontal extent of the field pattern from the microwave generator horn.   Since we have been interested in the use of phantoms, it seemed natural to extend this work by devising a simple method for the determination of field patterns with an absorbing plane.

Consequently, a 24 in. square sheet, 1/4 in. thick, of highly absorbing plastic was placed perpendicular to the axis of the microwave horn.   A grid was ruled on the sheet with 3 cm spacings, and at the junctions of the horizontal and vertical grid lines, small holes were drilled into which thermocouples were introduced.   A total of 20 thermocouples was then connected to a stepping switch from where each lead could be connected into a high gain dc amplifier and recorder.   The absorber sheet was then placed into a 30 in. cubed Lucite box in which the ambient temperature was carefully controlled.   A photograph of the arrangement is shown in Figures 1 and 2.

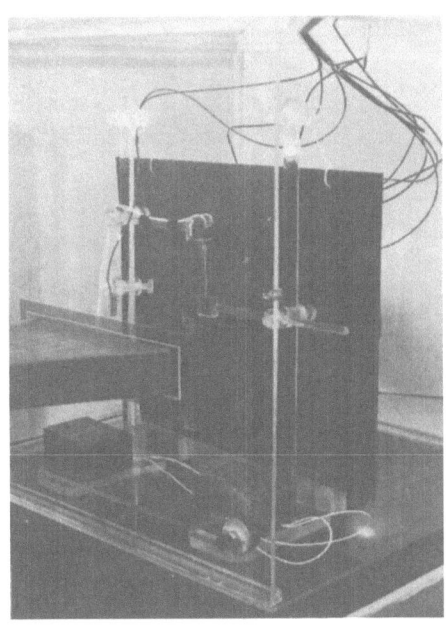

Fig. 1. Front view of field pattern absorber plate in Lucite box.

In use during the exposure to microwave power, the temperature gradients from all thermocouples were recorded alternately with 7½ sec for each measurement. The recorded data were then translated into heating rates in °C/min and plotted on a graph. Finally, iso-thermal curves were drawn which then represent the field pattern and the extent of the uniformity of the energy transfer from the microwave beam.

The simplicity of the method lies in the fact that only one single setup of equipment is needed and that the field remains uniformly disturbed in contrast to other types of field pattern measurements where the measuring device must be moved manually from point-to-point, and thus may represent a disturbing influence to the field pattern by virtue of the need for moving the detector horizontally and vertically.

Figure 3 shows one of the pattern plots as an example of the results of the method.

As an extension of this method, we plan to investigate the effect of curved absorber plates and thereby attempt to simulate, at least, approximately the effect of an animal's position in the microwave

FIG. 2.   Rear view of field pattern absorber plate showing thermocouples lead wires.

field with its possible change of the field pattern because of its irregular shape.

Another experiment, also in progress, is an attempt to investigate the reflection phenomena which seem to occur during the animal exposures. From observation of dogs during exposure in the microwave field, it was found that the animals were restlessly searching for a position in their cages which would offer them the greatest comfort. After finding such locations, the animals appeared more calm and would stay for longer periods without changing their positions. This phenomenon suggested that the animals had found a position where possibly the reflection of microwave power was at a maximum, and therefore, lower total absorption made them more comfortable. In order to test this observation, the following experiment was devised. A number of small calorimeter cups were built consisting of Lucite containers with 5 cc capacity into which water was introduced. Thermocouples measured the temperature rise when the calorimeter cups were placed into the microwave beam. Figure 4 shows the construction of the cups and their associated equipment, the thermocouple, and the small stirrer assembly. During the experiment, these cups were placed at various positions in front of the microwave horn

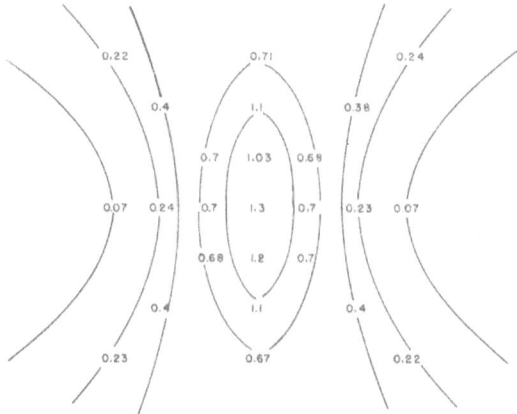

Fig. 3. Power absorption field pattern plot. Numerical
data refer to temperature rise rates in °C/min.

as shown in Figures 5a and 5b. They were exposed to controlled
power fluxes and their energy absorption was measured from temperature rise rates. Then a phantom was placed in such a position
that the front of the phantom was 1 cm from the back of the calorimeter cup situated in the center of the microwave beam. Again, the

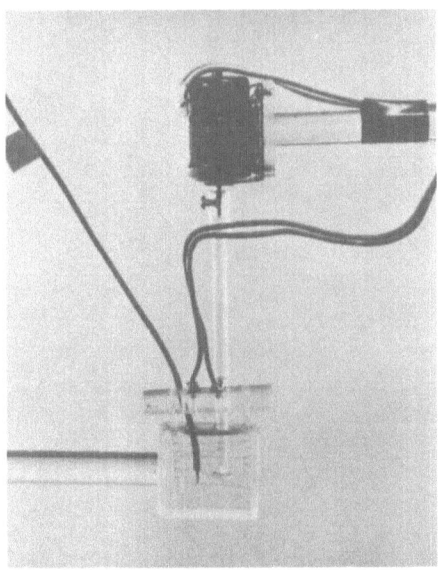

Fig. 4. Close-up of calorimeter
cup, stirrer, and heater coil.

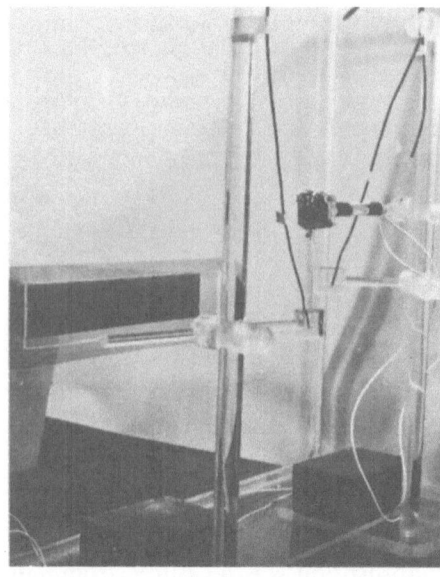

FIG. 5a. Position of calorimeter cups in front of microwave horn.

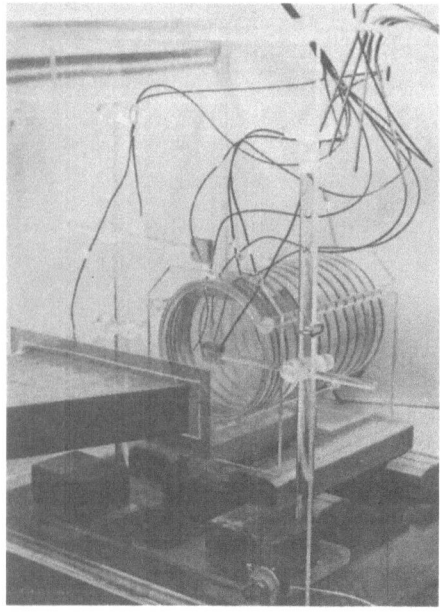

FIG. 5b. Position of calorimeter cups and phantom in front of horn.

temperature rise time was recorded and the energy absorption calculated. Several runs were performed with various distances of the calorimeter cups from the frontal surface of the phantom. The results are shown in Table I. From these data it can be inferred that with the phantom positioned at different distances from the horn, there exists obvious variations in the reflections and consequently absorption conditions. Measurements of the temperature rise within the phantom indicate similar changes as the distance of the phantom varies and these are obvious.

It appears from these experiments that animals indeed may find a particular position during their exposure periods where the reflection loss increases their apparent comfort. It seems rather obvious that the reported phantom and calorimeter studies are not at all complete and that further studies are needed to take into consideration such variables as curved and irregular surfaces as well as changes in the specimen positions within the Fresnel zone where most of the biological experiments are carried out.

TABLE I

**Heating Rates in Calorimeter Cups Without and
With Reflection from Phantom**

a) Without Phantom

Cal. Cup #	Temp. Rise °C/min
1	1.65
2	1.23
3	0.28

b) With Phantom

Cal. Cup #	Distance from Phantom	Temp. Rise °C/min
1	1 cm	4.25
1	3 cm	1.85
1	5 cm	2.25
2	1 cm	1.25
2	3 cm	1.33
2	5 cm	1.25
3	15 cm	0.35
3	16.5 cm	0.24
3	18 cm	0.1

Finally, I would like to report on the behavior and use of the calorimeter cups as a means of measuring the available microwave power for exposure of biological test specimen.

Since our group is interested in the absorption of microwave power rather than the incident flux, it should be more realistic to employ a unit of measurement commensurate with such power absorption. Therefore, if we measure w/cc in an absorber whose dielectric constant is similar to that of tissue and whose volume would represent a finite attenuation of microwave beams again similar to that of tissue, then a comparison might be achieved between experimental results from different investigators' laboratories.

The calorimeter cups mentioned above appear to have qualifications for such universally applicable volume power measurement. This, for instance, is an example: The cups used in our laboratory contain 5 cc of water, thus requiring 5 cal of heat energy to raise the temperature by 1°. Since they are equipped with small stirrers which achieved complete mixing of the calorimeter fluid, no complicated calculation or integration of heat diffusion gradients with varying depth in the calorimeter cup are needed. The heat rise time and the difference in temperature before and after exposure represent the principal parameters for interpretation of microwave power flux at a given point in space.

Absolute calibration of the device is also feasible by the well known substitution method by introducing an accurately known quantity of ac power into a small heater built into the calorimeter cup. Obviously, the method is not new; however, the simplicity of the device may offer some advantage in that it requires no tuning to any specific wavelength at least in the range from 1,000 to 10,000 mc. Within this range it has been calculated that the absorption power of water does not materially change.

An experimental device of this kind has been in use in our laboratory for some time now. During use of the calorimeter cup, there has been one particular precaution necessary. Since we have found it reasonably simple to measure temperature gradients of 0.01°/min with the associated amplifier and recorder, it has been found essential that the ambient temperature of the surroundings must be maintained constant. The arrangement of the apparatus is illustrated in Figures 6a and 6b. A Lucite box contains all measuring devices and the phantom or any other test objects. The room where the experiments are performed is air conditioned to $20 \pm 0.5°C$. The changes of temperature in the room and the box are continuously recorded

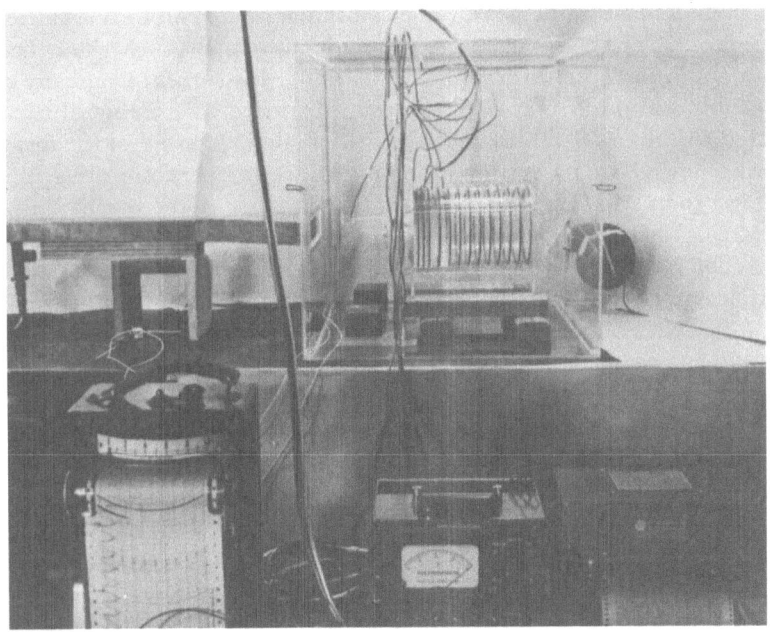

FIG. 6a. Arrangement of apparatus for measuring energy absorption from microwave near-field exposures of phantom.

FIG. 6b. Arrangement of apparatus for measuring energy absorption from microwave near-field exposures of phantom.

so that all measurements with the calorimeter cups can be normalized to account for any small changes in ambient temperature. The thermocouple currents are amplified by a Taylor Instrument Company dc amplifier and all information is then recorded by a Brown Instrument Company 0–50 mv recorder. By scale expansion of the measuring device, we have been able to measure temperature differences of less than 0.01°C. The device is thus capable of measuring differences of microwave power of the order of 0.1 mw.

The possible use of a more refined model of this type may permit an attempt at standardization of microwave dosimetry in other laboratories where biological effects of microwaves are studied.

In summary, a brief progress report is presented concerning some physical measurements as an aid to the biological investigators. Simplified methods of field pattern measurements, an attempt at quantitizing reflection losses, and a proposed simple power absorption measuring device have been discussed.

Relative Microwave Absorption Cross Sections of Biological Significance

A. Anne, M. Saito, O. M. Salati,
and H. P. Schwan
The Moore School of Electrical Engineering
University of Pennsylvania
Philadelphia, Pennsylvania

1. INTRODUCTION

A THEORETICAL INVESTIGATION was initiated to determine the relative absorption cross section of mankind to microwaves. This study resulted in the choice of appropriate phantoms which would simulate mankind for both theoretical calculations and measurements. Calculations and measurements were carried out to determine the relative absorption cross section of these phantoms which were filled with solutions representative of biological tissue at microwave frequencies. This work was in three phases:

(a) Theoretical calculations of the absorption cross section of spheres having complex dielectric constants of pertinence and correlation with experimental observations.

(b) Construction and final calibration of a microwave anechoic chamber for absorption and scatter studies. This also included the calibration of an "S" band radar used as a source of microwave power for the absorption studies.

(c) Experimental studies of absorption cross section of spheres filled with electrolyte mixtures simulating the dielectric properties of the human body.

Section 2 of this paper covers the basic approach to the problem of determining how much microwave energy is absorbed by a human being exposed to a microwave field. The approximations necessary in order to use phantoms rather than live subjects are covered in detail.

Section 3 discusses the theory of absorption of electromagnetic radiation by obstacles exposed to plane wave fields. The theory starts from classical solutions of Maxwell's equations, subject to appropriate boundary conditions. These solutions are modified for the case of spheres containing media of complex dielectric constants. A definition of relative absorption cross section is given.

The experimental set-up for checking the theoretical calculations is described in Section 4. It contains a description of the microwave anechoic chamber and its calibration, the microwave generator used for illuminating the samples of interest and overall calibrations and results.

2. THEORETICAL AND EXPERIMENTAL APPROACH

General

A decision was made during the early stages of this work to use phantoms rather than small animals to determine the relative absorption cross section of mankind to microwave fields. This decision was made because phantoms are stable, reproducible, and often subject to theoretical analysis. The behavior of animals involves many unknown factors and is presently not subject to theoretical analysis or extrapolation to mankind as has been explained by us on previous occasions.

In phantom studies one can make as exact a facsimile of the human body from an electrical point of view as is necessary to satisfactorily explore its scattering and absorbing properties.

Electrical Substitutes for Body Tissues

We have shown previously that electrical substitutes for body tissues in the microwave range are useful for all purposes where phantom studies, simulating the human body, are planned (1). All body tissues fall within the following range of electrical values throughout the total microwave frequency range:

Relative dielectric constant $\epsilon = 5$ to 70
Specific resistance $\rho = 10$ to $10{,}000$ ohm-cm
Relative permeability $\mu = 1$

Depending on frequency and type of tissue, a variety of probable combinations of dielectric constants and specific resistances within these ranges are possible. As previously reported (1), electrolyte mixtures having suitable combinations of properties in the above range were developed.

Microwave Considerations

In practical situations where mankind may be exposed to microwave fields, the electromagnetic wave falling on his body may vary from a homogeneous plane wave (at large distances from the source) to very complex waves close to the source. Plane wave fields were assumed in this work so that both theoretical and experimental results could be compared. It is hoped that the plane wave results may be extended to complex fields at some later date.

Even for plane wave fields, the theoretical problems are severe unless very simple objects such as homogeneous isotropic spheres or cylinders are chosen for the phantom. For experimental work, any shape may be used. If then, experimental and theoretical results show good agreement for simple shapes, there will be considerable confidence in experimental results for complex shapes such as the human body. In the present study, for both theoretical and experimental cases, the homogeneous isotropic sphere was used.

3. THEORY OF ABSORPTION OF ELECTROMAGNETIC RADIATION BY SPHERES HAVING COMPLEX DIELECTRIC CONSTANTS

Theory of Scattering

The first treatment of the problem of the scattering of a plane electromagnetic wave by a sphere of arbitrary size and electrical properties was given by Mie (2) in 1908. The present approach is to use the method of Stratton (3, Chap 9), who solved the problem by the use of orthogonal spherical vector wave functions.

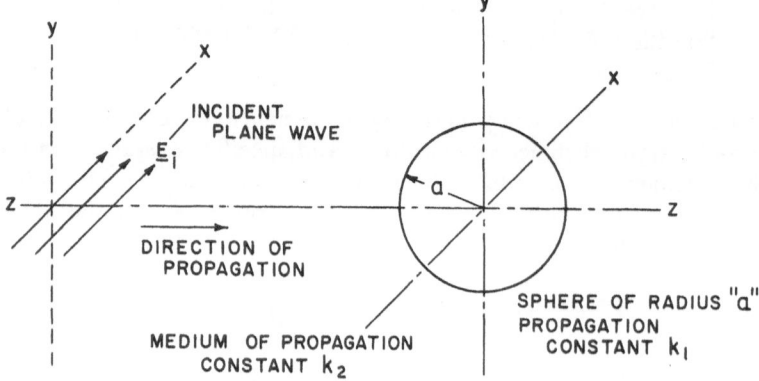

FIG. 1. Coordinates for a sphere in a plane wave field.

Let us assume that a homogeneous isotropic sphere of radius a, propagation constant k_1, and electromagnetic properties ϵ_1 (permittivity) and μ_1 (permeability) is embedded in an infinite, homogeneous, isotropic medium of propagation constant k_2 and properties ϵ_2, μ_2. A uniform plane wave, linearly polarized in the x-direction, is propagated through the medium in the positive z-direction (Fig. 1). According to Stratton (3, p. 569) eqs. 25 and 26, where eq. 26 should have a negative sign as shown below, the scattered energy is given by:

$$W_s = \pi(E_0/k_2^2)\sqrt{\epsilon_2/\mu_2}\sum_{n=1}^{\infty}(2n+1)\,(|a_n^r|^2 + |b_n^r|^2). \quad (1)$$

The total energy received by the sphere from the incident wave and then in part scattered and in part absorbed is given by:

$$W_t = -\pi(E_0/k_2^2)\sqrt{\epsilon_2/\mu_2}\,\Re_e\sum_{n=1}^{\infty}(2n+1)\,(a_n^r + b_n^r). \quad (2)$$

The coefficients a_n^r and b_n^r are, respectively, the amplitudes of oscillations of magnetic and electric type of the external field and are given by [(3), eqs. 10 and 11, p. 565; time dependence of $e^{i\omega t}$ is assumed here, instead of $e^{-i\omega t}$ assumed by Stratton]:

$$a_n^r = -\frac{\mu_1 j_n(N\alpha)\,[\alpha j_n(\alpha)]' - \mu_2 j_n(\alpha)\,[N\alpha j_n(N\alpha)]'}{\mu_1 j_n(N\alpha)\,[\alpha h_n^{(2)}(\alpha)]' - \mu_2 h_n^{(2)}(\alpha)\,[N\alpha j_n(N\alpha)]'} \quad (3)$$

$$b_n^r = -\frac{\mu_1 j_n(\alpha)\,[N\alpha j_n(N\alpha)]' - \mu_2 N^2 j_n(N\alpha)\,[\alpha j_n(\alpha)]'}{\mu_1 h_n^{(2)}(\alpha)\,[N\alpha j_n(N\alpha)]' - \mu_2 N^2 j_n(N\alpha)\,[\alpha h_n^{(2)}(\alpha)]'} \quad (4)$$

where $\qquad\qquad N = k_1/k_2 =$ index of refraction $\qquad\qquad$ (5)

$$\alpha = K_2 a \qquad\qquad\qquad (6)$$

$E_0 =$ amplitude of the incident field.

The primes on the square brackets in eqs. 3 and 4 indicate differentiation with respect to the argument of the Bessel functions j_n and $h_n^{(2)}$ inside the brackets. The index of refraction, N, may be real or complex. The functions $j_n(\alpha)$ and $h_n^{(2)}(\alpha)$ are the spherical Bessel and Hankel functions:

$$j_n(\alpha) = \sqrt{\pi/2\alpha}\, J_{n+1/2}(\alpha) \qquad\qquad (7)$$

$$h_n^{(2)} = \sqrt{\pi/2\alpha}\, H_{n+1/2}^{(2)}(\alpha). \qquad\qquad (8)$$

Assuming a time dependence of $e^{i\omega t}$ for the electric and magnetic field vectors and also assuming that sphere is a lossy dielectric and that it is embedded in an external medium which is air, then the propagation constants k_1 and k_2 become:

$$k_1 = \sqrt{\omega^2 \epsilon_1 \mu_1 - i\kappa\mu_1\omega} \qquad\qquad (9)$$

$$k_2 = \sqrt{\omega^2 \epsilon_2 \mu_2} = 2\pi/\lambda \qquad\qquad (10)$$

where: $\quad\omega =$ angular frequency of wave

$\qquad\lambda =$ wavelength, in meters, in free space

$\qquad\epsilon_1 =$ permittivity, farads/meter of the sphere

$\qquad\mu_1 =$ permeability, henries/meter of the sphere

$\qquad\kappa =$ conductivity, mho/meter of the sphere

$\qquad\epsilon_2 =$ permittivity of air

$\qquad\quad = 8.854 \times 10^{-12}$ farads/meter

$\qquad\mu_2 =$ permeability of air

$\qquad\quad = 4\pi \times 10^{-7}$ henries/meter

It has been previously stated that the relative permeability of tissues is unity and since the relative permeability of free space is also unity, $\mu_1 = \mu_2$ in the above equations. Thus eqs. 5 and 6 become:

$$N = k_1/k_2 = \sqrt{\epsilon - i(\kappa/\omega\epsilon_2)} \qquad\qquad (11)$$

$$\alpha = k_2 a = 2\pi(a/\lambda) \qquad\qquad (12)$$

where $i = \sqrt{-1}$, $\epsilon = \epsilon_1/\epsilon_2$ is the relative dielectric constant of the sphere, and $[\epsilon - i(\kappa/\omega\epsilon_2)]$ is the complex relative dielectric constant of the sphere.

The coefficients a_n^r and b_n^r can now be reduced to:

$$a_n^r = -\frac{j_n(N\alpha)\,[\alpha j_n(\alpha)]' - j_n(\alpha)\,[N\alpha j_n(N\alpha)]'}{j_n(N\alpha)\,[\alpha h_n^{(2)}(\alpha)]' - h_n^{(2)}(\alpha)\,[N\alpha j_n(N\alpha)]'} \tag{13}$$

$$b_n^r = -\frac{j_n(\alpha)\,[N\alpha j_n(N\alpha)]' - N^2 j_n(N\alpha)\,[\alpha j_n(\alpha)]'}{h_n^{(2)}(\alpha)\,[N\alpha j_n(N\alpha)]' - N^2 j_n(N\alpha)\,[\alpha h_n^{(2)}(\alpha)]'}. \tag{14}$$

Relative Absorption Cross Section

The power absorbed by a sphere of complex dielectric constant from a uniform plane wave field is given by W_a:

$$W_a = W_t - W_s \tag{15}$$

where W_t and W_s were previously defined in eqs. 2 and 1.

The relative absorption cross section S of the sphere is defined as the ratio of the absorbed energy per second to the power incident on the geometric cross-sectional area, $A = \pi a^2$, of the sphere ("shadow cross section").

The incident power is given by the product of the power density prior to the insertion of the sphere into the field (measured at the position of the sphere) and of the geometrical cross section, A, of the sphere. In free space, the power density P_d of a uniform plane wave is given by:

$$P_d = (E_0^2/2)\,\sqrt{\epsilon_2/\mu_2} = E_0^2/754 \text{ watts.} \tag{16}$$

The relative absorption cross section, S, is then:

$$\begin{aligned} S = W_a/P_d\pi a^2 &= 2W_a/E_0^2\pi a^2\,\sqrt{\epsilon_2/\mu_2} \\ &= (2/\alpha^2)\left[\sum_{n=1}^{\infty} -(2n+1)\,\{\Re_e(a_n^r + b_n^r) + (|a_n^r|^2 + |b_n^r|^2)\}\right]. \end{aligned} \tag{17}$$

The relative absorption cross section "S" is really an "efficiency factor." It gives a measure of the weight to be given to the geometric cross section of the sphere in order to obtain the absorption cross section.

Limiting Case: $\alpha \ll 1$

If the wavelength of the plane wave field is very large compared to the radius of the sphere so that $\alpha \ll 1$ (see eq. 12), the magnitudes

of the electric and magnetic oscillations of second order and above
are small compared with the magnitude of the first-order electric
oscillations (3, p. 570). Under these conditions, the relative absorption cross section depend only on the magnitude of the first-order
electric oscillations. The amplitude coefficient b_1^e of the electric
oscillation is given by (3, eq. 41, p. 572):

$$b_1^e = -(2i/3)\,[(N^2 - 1)/(N^2 + 2)]\,\alpha^3. \tag{18}$$

The relative absorption cross section now becomes:

$$
\begin{aligned}
S &= 12\,\frac{\kappa/\omega\in_2}{(2 + \in)^2 + (\kappa/\omega\in_2)^2}\,\alpha \\[2mm]
&= \frac{4524a\kappa}{(2 + \in)^2 + (\kappa/\omega\in_2)^2}.
\end{aligned}
\tag{19}
$$

Equation 19 for the relative absorption cross section is the same
as our eq. 9 in ref. 1, which was derived under the assumption that
the wavelength of the radiation is large compared with the radius of
the sphere. From the above equation, it can be seen that S is proportional to the radius of the sphere, and if the relative dielectric constant $\in \gg \kappa/\omega\in_2$, it is proportional to the conductivity κ.

Computations and Results

The above equations were slightly rearranged and programmed
for solution on a Univac Digital Computer (4).

The results of the machine computations are shown in graphical
form in Figures 2 through 7 as a function of α. The relative absorption cross section was calculated for $\in = 60$ and $\kappa = 1, 2$, and
10 mmhos/cm. The calculations were performed for values of α in
the following ranges:

$$
\begin{aligned}
\alpha = 0.01 &\text{ to } 2 \text{ in steps of } 0.01 \\
2.2 &\text{ to } 6 \text{ in steps of } 0.2 \\
7 &\text{ to } 24 \text{ in steps of } 1
\end{aligned}
$$

Looking over the data, several features of outstanding importance
can be enumerated.

(a) For small α, S increases linearly with the conductivity until
$\alpha = 0.3$ and increases linearly with α until $\alpha = 0.03$, both as stated
by eq. 19. (See Fig. 8.)

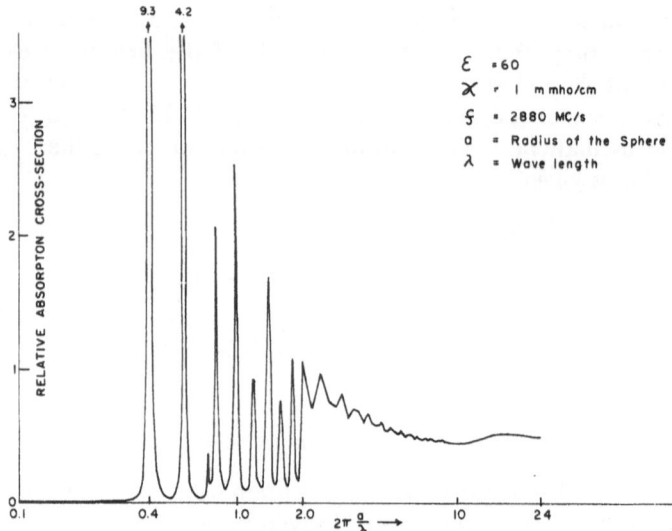

FIG. 2. Relative absorption cross section *S*.

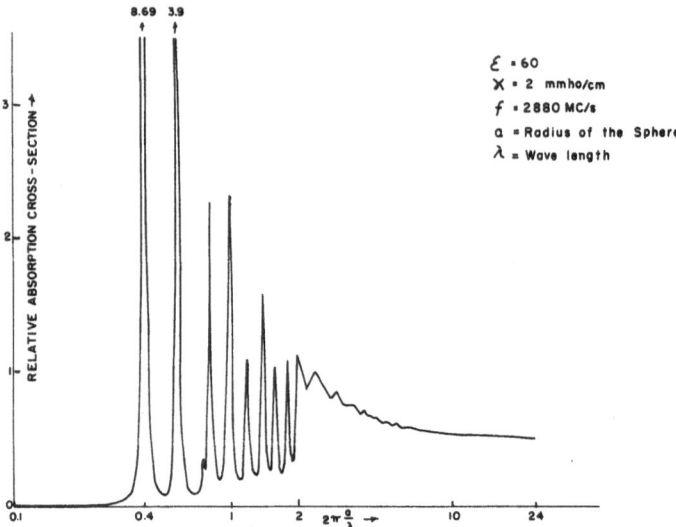

FIG. 3. Relative absorption cross section *S*.

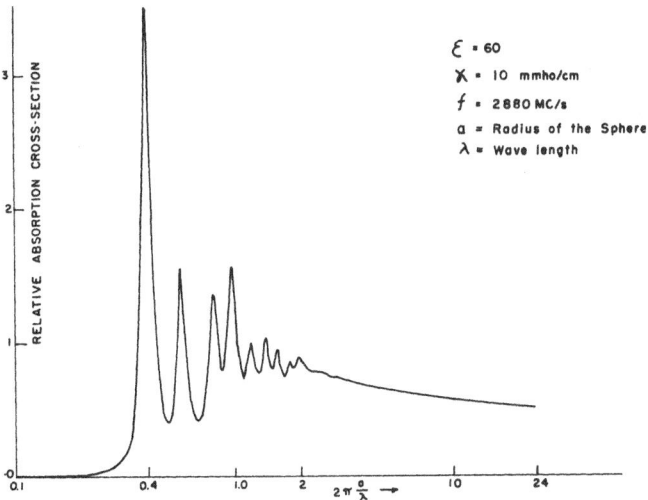

FIG. 4.   Relative absorption cross section S.

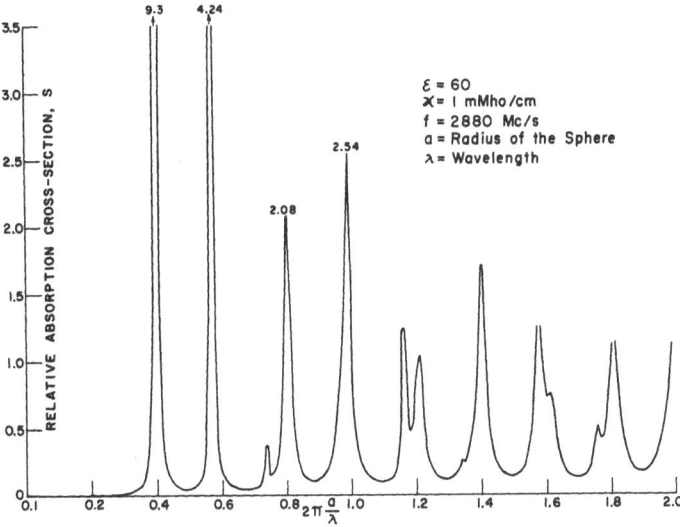

FIG. 5.   Relative absorption cross section S.

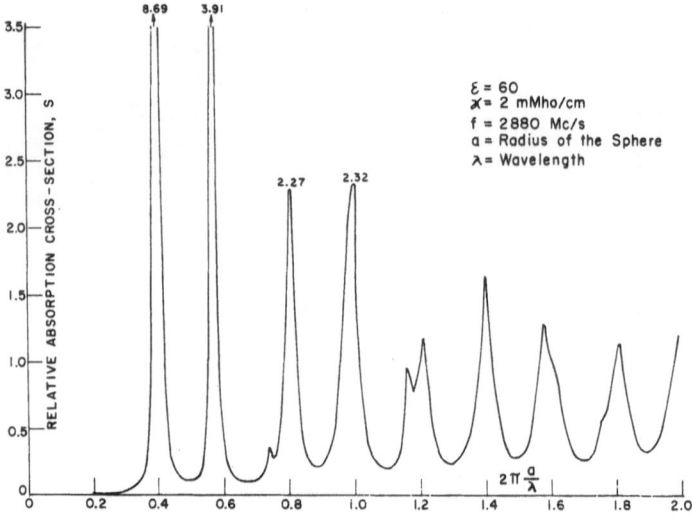

FIG. 6. Relative absorption cross section S.

(*b*) The first maximum value of S, for any conductivity, occurs at $\alpha = 0.4$.

(*c*) The S values at the first maximum decrease as the conductivity increases.

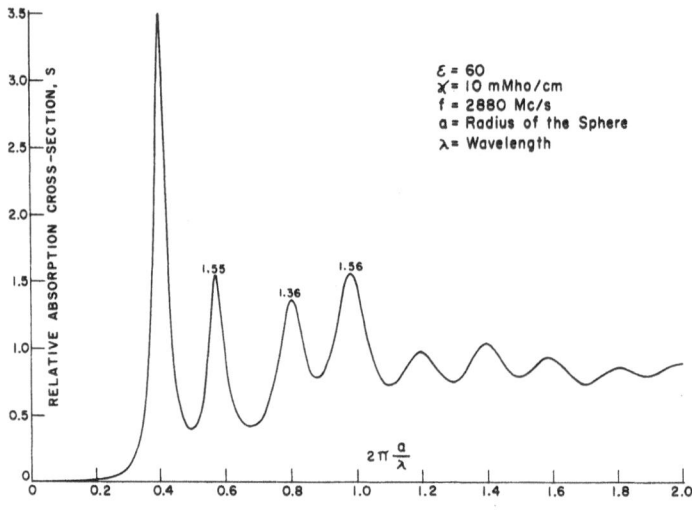

FIG. 7. Relative absorption cross section S.

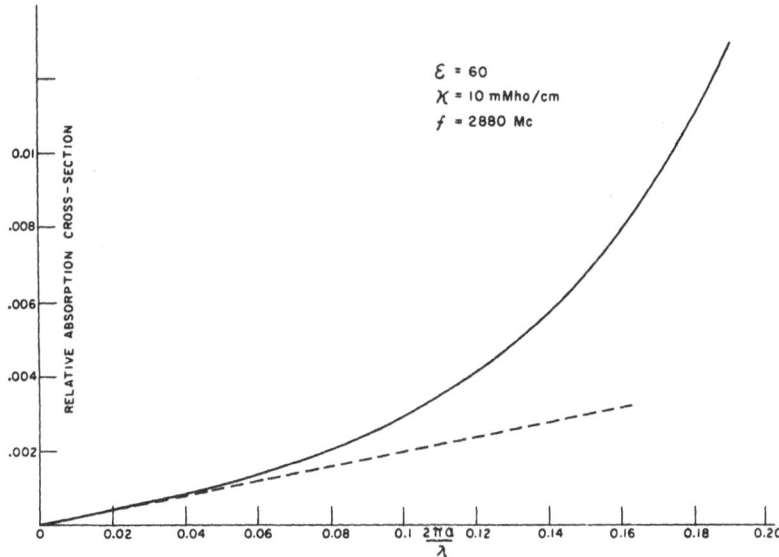

FIG. 8. Proportionality of S to a up to $\alpha = 0.03$.

(d) The relative absorption cross section S oscillates and the amplitudes of these oscillations decrease with increasing α. S is independent of the conductivity for large α, approaching asymptotically the value $S = 0.5$.

General Comments

The relative absorption cross section S is of great interest in assessing microwave damage to mankind. It can be seen from Figures 2 through 7 that it is a function of the conductivity of the exposed material as well as the size of the object relative to the wavelength of the incident field.

For objects much larger than the wavelength (α greater than 2.4), S remains below unity and decreases with increasing α. Below $\alpha = 2.4$, that is for objects small compared with the wavelength, S can become greater than unity. For this case, the object absorbs more power than is incident on its geometric cross section. If for instance, a part of the body, say the human head, is exposed to a plane wave microwave field at 400 mc/sec, the head may absorb 1.2 times the energy which is incident on its shadow cross section.

   This means that in conducting experimental work with objects which are small compared with the wavelength, small changes in object size or aspect may cause large changes in the absorbed energy.

   It can be seen from eqs. 13, 14, and 17 that S is a function of N, the index of refraction. From equation 11, N for fixed $\in$ is a function of the ratio κ/ω so that the results of the computations shown in Figures 2 through 7 can be extended to other values of κ and ω, providing the ratio κ/ω is held constant at one of the values shown on the figures. Thus the results of this report can be extended beyond the range of the data shown in the curves.

4. EXPERIMENTAL SET-UP

Anechoic Chamber and Antenna

   The anechoic chamber used in the experimental study is shown in Figure 9. Figure 10 is a photograph of the room showing the antenna and the phantom. The far wall of the room, facing the horn, is covered with 20 gage sheet copper to protect instruments and the occupants of the adjoining room. Three walls, the floor, and the ceiling of the room are covered with a single layer of Eccosorb 330 microwave material (the floor is covered with 330 FL so that it can be walked on). The side of the room near the antenna is arranged to have two movable walls of the same material so that dimensional changes are possible. The wall opposite the antenna, the one covered with copper, has a second layer of absorbing material arranged in a stepped or staggered fashion as shown in the figure. This treatment was found necessary in order to reduce reflections in the room and distortion of the transmitting antenna pattern (4).

   An exponential horn, Narda Model 644, having a gain of 15 db above an isotropic radiator, was used to illuminate the samples under test. The separation boundary between the Fresnel region and the far-zone fields for this antenna was at a distance of 5 ft from the mouth of the horn. Power density contours in a plane parallel to the mouth of the horn at a distance of 5 ft from the horn are shown in Figure 11. The contours are seen to be quite uniform. The criteria used in setting up the antenna in the room are given in ref. 5.

   From Figure 11 it can be seen that even for the largest sphere under test (22 liter, 34.3 cm diam), the variation in incident power density over the geometric cross section is less than 15%.

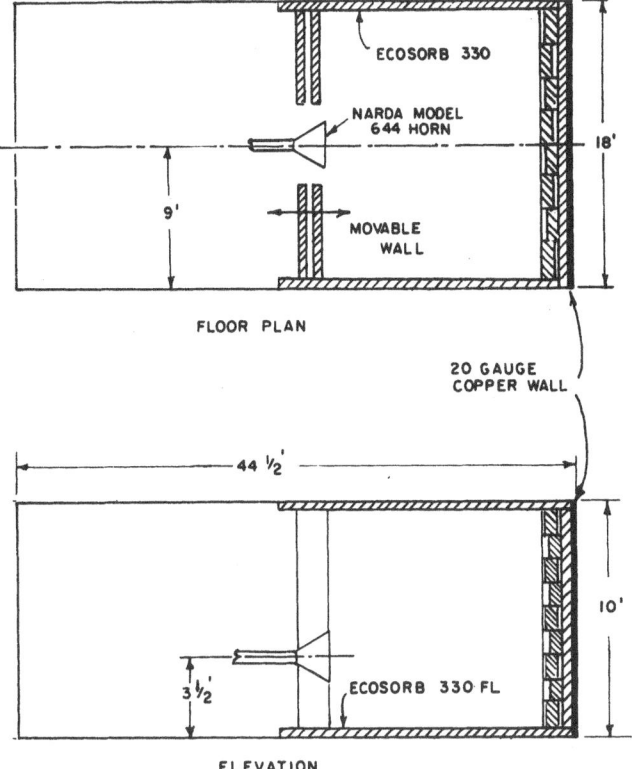

FIG. 9. Anechoic chamber.

The voltage standing wave ratio of the horn antenna in the anechoic chamber was measured with the slotted section and was found to be 1.03. All of the above measurements were made using a signal generator to feed the antenna.

Microwave Generator

An experimental model of the APS-20 radar was used as a source of microwave power for the experimental work.* The essential characteristics of this equipment are given below.

---

* This equipment was made available through the courtesy of the Office of Naval Research, Washington, D.C.

FIG. 10.   Photograph of anechoic chamber.

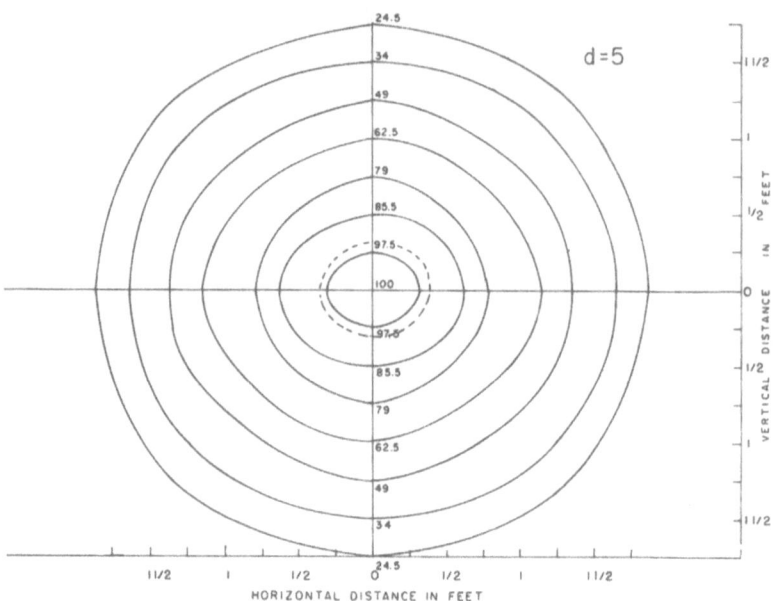

FIG. 11.   Power density contours in a plane parallel to the horn.

Frequency	2880 mc/sec
Wavelength	0.104 m
Peak power output	1.7×10^6 w
Average power output	1000 w
Pulse width	$2 \, \mu$ sec
Repetition rate	300/sec
Output system	RG-48/U "S" band waveguide

For calibration purposes, the output of the radar was connected to a water calorimeter load and to a Hewlett-Packard model 430 C power meter and model 477B thermistor mount via a directional coupler as shown in Figure 12.

The pulse shape and repetition rate of the radar were monitored via directional coupler B, a crystal detector, and an oscilloscope. The output power (average) was monitored via directional coupler A, attenuators, thermistor mount, and power meter. The average power output of the radar was measured by a water calorimeter using a substitution method. The accuracy of this measurement is about 5%. The voltage standing wave ratio of the load was measured with a slotted section and a standing wave meter and found to be 1.24.

The power output was measured and reference readings taken on the power meter after allowing about a half-hour for equipment warm-up. After this was completed, the water load was disconnected

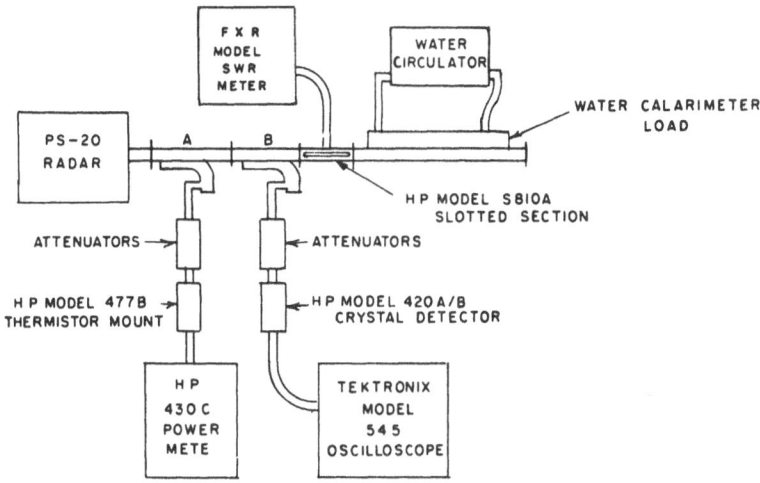

Fig. 12. Microwave generator calibration.

and the exponential horn antenna was connected to the system. In all measurements, the calibrated reading of the power meter was used as a measure of the power output.

With the antenna connected to the radar through two directional couplers used as a reflectometer, the voltage standing wave ratio was 1.188, corresponding to a power reflection coefficient of 0.0074, which indicates that almost all the power delivered to the antenna was radiated. Table I shows the power reflection coefficient and the corresponding voltage standing wave ratio for different sizes of the phantoms used. The results indicate that the effect of interaction between the antenna and the phantom is negligible, i.e., the experimental set-up is such that operation is in the far field of the antenna. The complete set-up as used in all further measurements is shown in Figure 13. Using the travelling carriage described in ref. 1, page 25, the power density at various distances on the axis of the horn was measured. The power densities in the far field of the antenna were calculated using the method described in ref. 5.

TABLE I

Effect of the Object on the Antenna Performance*

v	d = 5		d = 6	
	Γ	V.S.W.R.	Γ	V.S.W.R.
215	0.0074	1.188	0.0074	1.188
1050	0.0076	1.191	0.0073	1.187
2050	0.0073	1.187	0.0073	1.187
5150	0.0085	1.203	0.0082	1.199
12435	0.0082	1.199	0.0071	1.183
21150	0.0071	1.183	0.0066	1.176

d	v = 215		v = 5150		v = 21150	
	Γ	V.S.W.R.	Γ	V.S.W.R.	Γ	V.S.W.R.
4	0.0078	1.193	0.0096	1.218	0.0087	1.206
3	0.0082	1.199	0.0058	1.164	0.0083	1.201
2	0.0088	1.207	0.0008	1.059	0.0019	1.091
1	0.0103	1.226	not measured		not measured	

* d = distance between the object and the horn in ft. v = volume of the sphere in cc. Γ = reflected power/incident power = power reflection coefficient.

FIG. 13. Photograph of microwave generator set-up.

The measured and the calculated values of power density at various distances on the axis of the horn were given in Table II and found to agree rather closely.

The power densities in the Fresnel region of the antenna were calculated using the methods described in ref. 6 and were used in determining the relative absorption cross sections.

TABLE II

Comparison of Measured and Calculated Values of Power Density at Various Distances on the Axis of the Horn*

Distance from the horn, d, ft	Power density, mw/cm²	
	Measured	Calculated
5	56.7	55.7
6	40.2	38.7
7	31.0	28.4
8	22.3	21.7
9	15.8	17.2

* Transmitted power (P_t) $=$ 510 W.

Thermal Considerations and Phantoms

Hollow, spherical flasks of Pyrex glass of the following sizes were used in the experiments:

Capacity, liters	Diam, cm
0.215	7.42
1.05	12.61
2.05	15.76
5.15	21.42
12.44	28.74
21.15	34.31

The wall thickness of the flasks was less than ⅛ wavelength and is believed to be inconsequential in this work in view of the comparatively low dielectric losses of glass.

The flasks were placed in square boxes made of polystyrene foam ($\epsilon \doteq 1.02$), and air spaces between box and flask loosely packed with polyfoam granules (Fig. 14). The polyfoam was used as thermal insulation to cut down heat losses due to conduction and convection. It has negligible effect on the microwave properties of the phantoms.

The flasks were filled with the electrolyte mixture of interest and then placed in the anechoic chamber at various distances from the horn antenna (along the antenna axis).

The relative absorption cross section was calculated by measuring the temperature rise in a known time due to exposure of the phantom to the microwave field.

The heat developed in a homogeneous isotropic sphere placed in a uniform plane wave electromagnetic field is:

$$H = (4/3)\,\pi a^3 hm T \text{ calories} \tag{20}$$

where a = radius of sphere, cm
            T = temperature rise (°C) of the electrolyte
            h = specific heat of the electrolyte (cal/g/°C)
            m = specific weight of the electrolyte (g/cc)

and provided that heat losses to the outside can be neglected (see later).

The energy absorbed per second is then:

$$W_a = 4.2H/t \text{ watts} \tag{21}$$

where t = time in seconds.

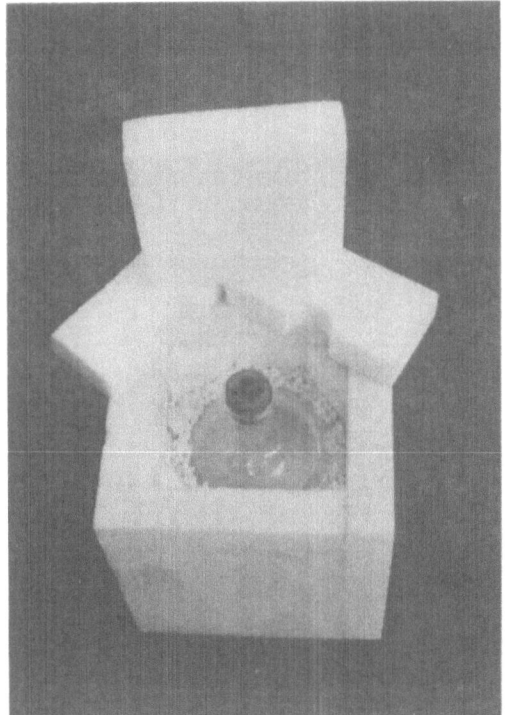

FIG. 14. Photograph of phantom.

The power incident on the geometrical cross section of the sphere is:

$$W_i = \pi a^2 P \text{ watts} \qquad (22)$$

where P = power density of field in w/cm^2.

The relative absorption cross section is then:

$$S = 5.6 hm Ta/Pt. \qquad (23)$$

Conduction and convection losses are negligible and eq. 20 applicable, if operation does not exceed the "linear portion" of the thermal transient (see Fig. 15). Consider, for example, a 5150-cc flask of radius $a = 10.7$ cm and filled with a saline solution having a dielectric constant $\epsilon = 78$ and a low frequency conductivity of $\kappa_0 = 9$ mmho/cm. The sphere was exposed to a microwave field of 93 mw/cm^2, and the average temperature rise of the electrolyte measured as a function of time. The temperature was measured

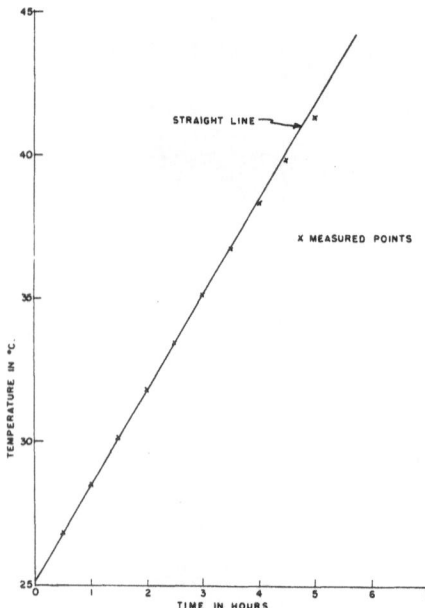

Fig. 15. Temperature characteristic of the measuring system.

after the solution had been stirred. The result is shown in Figure 15.
It can be seen that the temperature rise is a linear function of time
until the difference between the ambient and sample temperature is
about 15°C corresponding to an exposure time of about 4 hr.

Results

The relative absorption cross section was measured as a function
of the distance from the mouth of the horn as shown in Figure 16.
The relative absorption cross section is independent of the distance
from the horn from 3 to 12 ft. The deviation beyond 12 ft is perhaps
due to the presence of residual standing waves in the field near the
back wall of the room. Uncertainties due to the rapidly decreasing
resolution of the thermal technique as a consequence of the weakness
of the field far from the antenna further complicates this problem at
large distances. All subsequent work was done at distances of 5 and
6 ft from the horn where reasonable temperature rises were obtaina-
ble and where standing waves from room reflections were a minimum
and near field effects were essentially negligible.

Solutions of KCl-Dioxan-Water mixtures were then used in the phantom to simulate the electrical properties of human tissue. The dielectric constant was 60 and the low frequency conductivity, κ_0, 10 mmho/cm, which simulates closely all soft tissues of high water content. The relative absorption cross section for different size spheres was then determined. Table III shows the comparison of the experimental and theoretical results. A minor difference between experimental and theoretical data of about 5% is readily explained by the fact that the microwave conductivity κ of the test solution is not identical with its low frequency conductivity κ_0. The following equation holds as may be derived from ref. 1, eq. 1:

$$\kappa = \kappa_0 + 800(f/f_0)^2 = 30 \text{ mmho/cm}. \qquad (24)$$

The values obtained experimentally for 30 mmho/cm are compared with the theoretical values for the sphere sizes tested (Table III). They are seen to be a reasonable extrapolation of the theoretical values obtained at lower κ-values.

5. DISCUSSION AND SIGNIFICANCE

(1) The relative absorption cross section is near 0.5 for large values of α. A value of 0.5 ± 0.1 is approached for $\alpha = 6$, i.e., if the

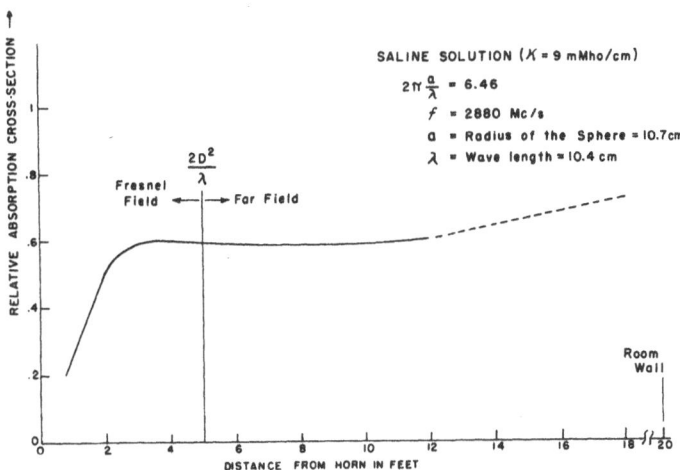

FIG. 16. Relative absorption cross section variation as a function of distance from the horn.

TABLE III

Relative Absorption Cross Sections of Spheres of KCl-Dioxan-Water Mixture*

Radius of the sphere	α	Relative Absorption Cross Section			
		Theoretical			Experimental
		$\kappa = 1$	$\kappa = 2$	$\kappa = 10$	$\kappa = 30$ mmho/cm
3.71	2.24	0.72	0.87	0.79	0.79
6.30	3.80	0.61	0.69	0.68	0.71
7.88	4.75	0.55	0.63	0.65	0.67
10.74	6.46	0.51	0.58	0.61	0.64
14.37	8.67	0.47	0.55	0.58	0.63
17.15	10.35	0.44	0.54	0.56	0.59

* Dielectric constant = 60. Frequency = 2880 mc/sec.

radius "a" compares with the wavelength. This is true, independently of the value of κ within the range of κ-values investigated.

(2) While the calculations and experiments reported above have been performed for a frequency of nearly 3000 mc, they can be extended to lower frequencies. From eqs. 17, 3, and 4:

$$S = f(a/\lambda, \; \epsilon - i\kappa/\omega\epsilon_2). \tag{25}$$

This states that the relative absorption cross section is merely a function of the *ratio* a/λ, i.e., of curvature and wavelength and of the complex dielectric constant. Thus it is possible to increase wavelength and radius by the same factor and obtain the same relative cross section provided that the complex dielectric constant is also kept the same. This latter condition merely requires proportional adjustment of the conductivity with the frequency. It is fortunate that the dielectric constant of most tissues of high water content is practically frequency independent from 300 to 3000 mc. Thus the data given are not only useful for 2880 mc, but also for other frequencies provided that conductivities are simply proportionally adjusted with the change in frequency. Of course the question arises as to whether these adjusted conductivity values are biologically significant, i.e., identical with those observed for the soft tissues of primary interest here. In examining latter values [see Schwan (7)] it appears that the combinations 3000 mc (30 mmho/cm) and 1000 mc (10 mmho/cm) are particularly biologically significant. This indicates that the absorption cross section data, which have been experimentally obtained are significant for both 1000 and 3000 mc. This implies a considerably broader applicability of the stated cross sections.

(3) The near independence of the relative absorption cross section data S from κ for a given ratio $\alpha = 2\pi a/\lambda > 6$ is identical with a corresponding independence of S from ω, making the data applicable to a major part of the frequency range of biological interest.

(4) The independence of the absorption cross section from radius for values of α larger than 6 implies that curvature does not substantially affect absorption characteristics, at least as long as the radius of curvature is larger than the wavelength. Thus it appears that structures whose cross section is larger than $\pi\lambda^2$ have a relative absorption of 50%, independent of their shape. This holds provided that no substantial contributions to absorption can result from local structures of rapid curvature, as characterized by a radius much smaller than the wavelength. This statement applies particularly to mankind, since its cross section is almost $1 \text{ m}^2 = 10,000 \text{ cm}^2$ for wavelength values smaller than about 60 cm, i.e., frequencies above 500 mc. High local possible absorption by structures, such as the nose or the ears, should not contribute to the total absorption noticeably in view of the small local volume involved. This does not mean, however, that these parts may not be damaged seriously while the body is affected negligibly.

6. SUMMARY

The relative absorption cross section of mankind is near 50% at 3000 mc. It does not vary strongly with frequency for frequencies above 300 mc.

References

1. Schwan, H. et al., "Effects of Microwaves on Mankind," *Second Ann. Prog. Rept. on Contract AF41(657)129,* Univ. of Penn., 1 March 1959.

2. Mie, G., *Ann. Physik, 25,* 377 (1908).

3. Stratton, J. A., *Electromagnetic Theory,* McGraw-Hill, New York, 1941.

4. Schwan, H., et al., "Effects of Microwaves on Mankind," *Third Ann. Prog. Rept. on Contract AF41(657)129,* Univ. of Penn., 31 July 1960.

5. Salati, O. M., "Microwave Absorption Measurements," Presented at the Microwave Investigators Conference, Patrick Air Force Base, 14, 15 January 1959.

6. Jacobs, E., "Fresnel Region Patterns and Gain Corrections of Large Rectangular Antennas," *Proc. Fifth Conf. on Radio Frequency Reduction*, Armour Research Foundation, Chicago, Ill., October 1959, p. 499.

7. Schwan, H. P., *Medical Physics*, Vol. 3, Otto Glasser, ed., Year Book Publishers, Chicago, Ill., 1960. Schwan, H. P., and Li, K., "Capacity and Conductivity of Body Tissues at Ultrahigh Frequencies," *Proc. I.R.E., 41,* 1735 (Dec. 1953).

8. Aden, A. L., "Electromagnetic Scattering from Spheres with Sizes Comparable to the Wavelength," *J. Appl. Phys., 22,* 601 (May 1951).

9. Penndorf, R. B., "New Tables of Mie Scattering Functions for Spherical Particles," *Part 6, Geophysical Research Papers, No. 45,* Air Force Cambridge Research Center, 1956.

Biological Effects of Microwave Energy at 200 mc

C. H. Addington, C. Osborn, G. Swartz,
F. P. Fischer, R. A. Neubauer,
and Y. T. Sarkees
Departments of Biology and Electrical Engineering
The University of Buffalo
Buffalo, New York

INTRODUCTION

Work at the University of Buffalo is at 200 mc frequency using a 6 kw RCA transmitter, continuous wave, horn antenna with a launching loop oriented in such a manner as to make it possible to change the plane of polarity of the field. All experiments are carried out in a large anechoic chamber lined with appropriate echosorb and shielded at the surface with sheet metal. More complete descriptions of the equipment will be found in previous reports contained in the proceedings of the First, Second, and Third Annual Tri-Service Microwave Conferences.

During the months since our last reports to the Tri-Service Conference our experimental work has been directed toward the following problems:

1. Further development and improvement of miniaturized instrumentation which will operate with power on in a 200 mc field.

2. Irradiation of bacteria in an attempt to demonstrate possible nonthermal effects.

3. Study of the effects of chronic irradiation on guinea pigs.

4. Relate irradiation dosage to survival in an attempt to establish LD 50 values in dogs and guinea pigs. For the sake of brevity in this particular report, we emphasize primarily the data on our experiments which relate dosage to survival.

177

RECENT EXPERIMENTS

Dogs—Temperature Elevation Related to
Polarity of rf Field

In these experiments 28 dogs were used ranging in weight from 24 to 69 lb and averaging 44.8 lb. Some of the animals were irradiated as many as 9 times, always allowing 3 or 4 weeks to elapse between exposures. A few of the animals were exposed only once. On the average, each animal was exposed about 3 times. Temperatures were taken in some instances intermittently during the progress of the experiment with power off, using a rectal thermometer. In other instances temperatures were taken at the end of the exposure to avoid interruptions during the exposure period. As will be seen in Table I, animals were tested at 8 different field intensities, only 6 of which were of sufficient strength to elevate the temperature more than 3°F in a period of 1 hr or less. In Table I we may focus our attention on the 6 levels of intensity ranging from 38 mw/cm² to a

TABLE I

Minutes to Raise Rectal Temperature Related to Polarity*

mw/cm², 200 mc	°F elevation			
	3	5	7	9
330	4.5 (?)	7.2 (?)	12.0 (?)	14.0 (?)
220	7.1 (3.0)	12.7 (5.0)	21.0 (6.0)	32.0 (12.0)
194	10.2 (5.2)	17.1 (8.0)	27.0 (10.5)	
165	11.7 (8.5)	21.4 (10.9)	29.0 (12.1)	
105	21.0 (11.0)	26.0 (19.5)	—	
38	23.5 (18.7)	31.0 (26.0)	—	

* 10 and 22 mw intensities did not produce 3°F elevation in 60 min. Parenthetical values are from animals oriented in the Y axis; other values, X axis.

maximum of 330 mw/cm^2. At all but the highest intensity, animals were oriented in the X axis, at right angles to the plane of polarization, and in the Y axis, parallel to the plane of polarization. Change in polarity was accomplished by rotating the launching loop 90° rather than shifting the position of the animal as such. Table I represents the time required to elevate rectal temperature 3, 5, 7, or 9° as indicated in the respective columns.

Attention is invited to three observations: (1) At the low intensities of 10 and 22 mw, temperature elevations during a 60 min exposure were less than 3°. (2) Rate of temperature elevation as measured rectally increases as the intensity of the field increases. (3) When the long axis of the animal's body is parallel to the launching plane (Y axis) temperatures are elevated much more rapidly than when the animal's body is oriented in the opposite axis. Attention has been directed to this last observation in previous publications, but at that time only limited quantitative data were available to support it. We believe this "antenna effect" to be of considerable importance and suggest that it should not be lost sight of in field operations.

Exposure (Intensity and Duration) Related to Survival

Keeping the role of polarity in mind as an important factor in considering exposure in its relation to survival, let us turn next to experiments which relate dosage to temperature elevation and survival. These experiments used those animals referred to in our previous work plus 9 others. The weight of the additional animals fell within the range previously described and had no significant influence upon the average weight figure cited previously. The experiments to be described were carried out with the following objectives in mind:

1. To try to establish time-intensity dosages critical to survival.

2. To relate temperature elevations and animal responses with the polarity of the field.

3. To record the responses of dogs during the radiation (moving pictures were taken during two experiments).

4. To obtain autopsy data on those animals not surviving the radiation in an attempt to determine the cause of death or to help explain symptoms among survivors and possibly to throw some light upon suspected "hot spots" developing in animals during irradiation.

In the experiments which follow we have tried to be economical in our use of animal materials. We have attempted to identify advanced severe symptoms, such as profuse salivation, prostration, and coma, in an attempt to stop the exposure just before irreversible changes might set in. In this way we have held our mortality rate to a relatively low percentage (only 7 dogs have been lost in this series) although we have used a rather large number of dogs and have completed 4 or more experiments on each point in Table II. All of the values to be seen in Table II are averages and in some instances represent as many as 9 experiments. Because these are averages it will be found that the values do not agree in some cases with the related data presented in other tables.

Fewer experiments are represented by the data on the right side of Table II. We have reserved these experiments for the later phases of our work because we felt that additional experience gained with exposures in the X axis would be desirable before proceeding with more experiments in higher fields of intensity using the Y axis. These experiments are more difficult to do because, as will be seen, temperature elevations occur much more rapidly and the chance of a casualty with little or no warning is greater. Experiments to supplement these data are being done at the present time and will be completed shortly.

Turning now to the data presented in Table II we may make the following observations: Field intensities referring to work in the

TABLE II

Exposure (Intensity and Duration) Related to Survival

mw/cm², 200 mc	X axis			Y axis		
	Min.	°FΔ	% died	Min.	°FΔ	% died
330	15	9.0	50			
220	21	7.4	25	7	8.2	0
194	23	6.3	28	13	8.4	0
165	29	5.9	0	18	7.6	16
105	35	4.9	0	24	5.8	0
38	39	4.7	0	33	5.6	0
22	60	1.4	0			
10	60	0.4	0	50	1.6	0

X axis of less than 105 mw produced rectal temperature elevations of less than 5° within the exposure periods indicated. From our experience, we feel that a 5° rectal temperature rise is not excessive if not maintained over a long period of time and probably little or no damage results to the animal unless there are local "hot spots" which may be developing unknown to us and which would go unnoticed so long as rectal temperatures only are recorded. There will be more to report about this problem in connection with autopsy findings. At 194 mw an average 23 min exposure produced an average elevation in rectal temperatures of 6.3°F. Of the 7 animals exposed at this level, the 2 which died had rectal temperature elevations in excess of 7.5°. To date, none of our casualties have shown a rectal temperature elevation of less than 7.5°F. At 220 mw 2 of 8 animals succumbed, while at the highest intensity of 330 mw, 2 of 4 failed to survive. Among the data under the Y axis experiments in Table II, it will be noted that 1 of 6 animals receiving 165 mw/cm² died.

In attempting to attach L.D. 50 values to such data, not only must orientation of the animal in the field be noted but two other important variables, time and field intensity, must be considered. On the basis of our experience, we feel that the major consideration which determines whether the animal will live or die is the height to which the temperature goes and how long it stays there. An animal's temperature may be elevated suddenly by 9 or 10° or perhaps even more, and yet survival is possible if cooling occurs rapidly after a relatively short sojourn at the high temperature. On the other hand, an animal may succumb whose maximum temperature elevation has been only 7 or 8°F if it stays at that level for several (we estimate 5 to 10) min. Accordingly, it is safe to conclude that the percentage of lethality would increase at all of the three higher intensities if the length of exposure time were to be increased. Undoubtedly a casualty would even occur now and then at the 165 mw level with increased exposure time, although it will be seen from the data presented that the exposure times at 165 mw are approximately double those of the highest field intensity (330 mw). If the animal is oriented in the Y axis it is likewise safe to predict that casualties will occur at somewhat lower field intensities (as recorded by present methods) than in the X axis, and that at higher field intensities the length of the exposure time required to bring about death will be significantly lessened.

Guinea Pigs: Exposure (Intensity and Duration)
Related to Survival

Table III presents data on a total of 54 guinea pigs averaging 1 lb 14 oz in weight exposed at relatively high field intensities (300 to more than 600 mw/cm^2) for relatively short periods of time (10, 15, or 20 min). Values on each line represent 3 or more animals. Data such as these are somewhat disconcerting, even to a biologist, for they do not seem to present any consistent trend. It will be observed that deaths occurred at each of the intensities employed. In all, 24 casualties were recorded. Most of the deaths occurred within 2 hr after the termination of the exposure but in 3 or 4 instances the animals lingered on overnight or longer. In one instance, an animal expired 3 days after the experiment. During the exposure all of these animals were housed in a plastic box with widely spaced rods making up the walls. There was sufficient room in the box to allow the animal to turn around freely and orient his body in any position he chose. Notes taken during the irradiations indicate that some animals were much more placid than others and moved only a slight amount during the entire exposure. Mortality seems to be significantly lower among those animals which were most restless and moved about rather vigorously. This is not subject to quantitative

TABLE III

Guinea Pigs, Dosage Related to Survival

mw/cm^2	Min.	Ave. temp. elevation, °F	% mortality
330	20	7.3	100
	15	8.2	0
	10	2.2	0
410	20	7.5	67
	15	7.1	33
505	20	5.9	50
	15	7.2	33
570	20	6.8	0
	15	5.7	33
590	20	10.6	100
	15	5.4	50
635	15	8.6	100
	10	6.0	0

analysis but if there is virtue in this observation, it might support the idea that differential heating or "hot spots" may result in serious damage particularly in those animals which remain still several minutes at a time. These data are too limited to allow one to establish an L.D. 50 with certainty, but they do suggest that such a value might fall in the neighborhood of 400 to 500 mw for a 20 min period.

Autopsy Data

Rather detailed autopsies were done on those animals which succumbed during the experiment or directly following the exposure. The abdominal wall was opened and thermometers were plunged into several locations in and among the viscera. Of the data recorded the following observations may be of greatest interest.

In dog #17, whose rectal temperature reached a maximum of 113°, temperatures were recorded in the following locations: (a) midline abdomen just below diaphragm, 114°; (b) deep midline between diaphragm and liver, 116°; (c) inside stomach, 116°.

Temperatures of approximately 114° were recorded in several other intra-abdominal and intra-thoracic positions. These temperatures were recorded about 12 min after irradiation stopped. While no significance can be attached to the absolute values, they do indicate that extremely high temperatures are reached in the abdomen. Qualitative changes of particular interest were largely related to the gastro-intestinal tract. The greater curvature of the stomach was extensively inflamed and hemorrhagic. The jejunum and ilium were extensively inflamed and hemorrhagic blebs appeared intermittently. The anterior colon was distended but not extensively inflamed. The middle one-third was extensively inflamed while the posterior portion was only slightly inflamed. Hemorrhagic blebs which appeared to be the result of capillary or arteriolar breakdown were present in rather large numbers along the walls of the colon and small intestine. Other organs appeared essentially normal on superficial observation and when macroscopically observed in section.

In the majority of instances animals which have been subjected to high temperatures by irradiation show digestive upset usually in the form of extensive diarrhea during the hours subsequent to exposure. Such gastro-intestinal symptoms are consistent with the evidence of damage seen at autopsy and would seem to indicate that disproportionately high temperatures may be developing in certain parts of the body.

(Editor's note: A moving picture film of about 8 min length was shown at this point. It demonstrated some of the typical responses which are seen when a dog is exposed to irradiation at 200 mc at a field intensity of approximately 330 mw.)

IMPROVED BIOELECTRIC INSTRUMENTATION AT 200 mc

Field Strength

The most satisfactory laboratory method developed so far to measure components of electric field strengths at frequencies below the microwave region at almost point locations in the near and far field of an energized antenna is as follows: a miniature exploring and sensing circuit, without benefit of connecting wires to a remote read-out, possesses a memory that can be interrogated when the miniature measuring circuit is removed from the field. The present field strength capsule is designed to sample the peak value of a pulsed wave or of cw, but not the average value.

A shielded IN463 silicon diode is placed in series with a high grade memory capacitor in series with a folded sensing dipole which is immersed in the field. The capacitor voltage subsequently can be read from suitable terminals (by means of an electronic electrometer) when the entire compact unit is removed from the field. The back resistance of the diode is adequate to maintain the charge sufficiently long for sampling. If an average power detecting circuit is employed (with shunt diode), a thread-tripped or otherwise remotely controlled transistor-type microswitch is included in series with the memory in order to maintain its charge.

Loops can be employed instead of dipoles to measure the magnetic components of the electromagnetic field. Appropriate calibration is effected in the far field against more standard instrumentation.

Implantable Telemeter for Temperature Measurement under Influence of rf Field

Perhaps the most important instrument needed in the fundamental study of large living biological specimens subjected to intense sub-microwave frequencies of electromagnetic radiation is a non-disturbing microminiature thermometer which can report deep temperatures from a number of stations *while the power is applied.* It would seem that such instruments are now nearing realization.

Considerable effort is being expended in the U. S. on microminiaturization and packaging of electronic components. With each new advance, different types of circuitry become feasible. It is now believed that sufficient advances have been made to justify an actively energized implantable temperature telemeter capsule employing sonic FM transmission to a miniature transistor repeater station on the outside skin surface and thence to a more remote frequency measuring circuit.

Although no implants in living specimens have yet been made, prototypes of the telemeter have been constructed which when immersed in nonliving protein media displayed quite satisfactory miniaturization, life of power supply, and stability of calibration. Recording for the present has been accomplished with a commercial sound level meter with a cycle counter output.

A parallel group of 1.3 v miniature mercury cells in series with a hearing aid speaker (through a UTC DO-T6 impedance matching transformer) resides in the collector-base circuit of a 2N5 36 pnp transistor. The base-collector circuit is also coupled to the simple base-emitter circuit by means of a UTC DO-T1 transformer. There are no other elements in this blocking oscillator. The entire telemeter prototype can be packaged now in a stainless steel cylinder about ¾ in. in diameter and 1-in. long. The largest portion of the telemeter is the speaker plus its matching transformer. It is felt that this latter combination can be made smaller, perhaps to one-half size by more direct design.

The temperature sensitivity of the lone transistor creates most of the frequency deviations of the telemeter. At 100°F and 1.3 v of drive, one prototype oscillated at 400 cps with about 20 cps deviation/°F at nominal temperature.

Passive capsules, self-powered by the rf field, have been considered but because of the variable nature of (1) the region of exposure, (2) field amplitude, (3) pulse or cw operation, (4) movement of the animals, they were temporarily discarded. A space-taking voltage regulating circuit would also be required, but an advantage, if the unit were successful, would be an infinite life power supply.

A low level acoustical telemeter was chosen because it would have a useful life of several weeks and the acoustical method of transmission would require no electrical antenna or coupling coil to interfere with the rf field. Sustained spurious noise, e.g., from fans, must be eliminated to prevent a false reading.

A 2-transistor multivibrator was originally employed to drive the speaker but it required more components and occupied too much space. Also, if commercial microminiaturized multivibrators were employed to diminish the volume, the acoustical frequency exceeded the range of the speaker transducer. Magnetostriction-type transducers were not tried.

Externally located miniature transistor oscillators have been built which would measure skin temperature and telemeter this information to a photoelectric pickup by means of a flashing Mite-T-Lite (Sylvania Lighting Products) in a somewhat darkened room.

SUMMARY

A memory method for measuring the components of the field-strength vectors at an almost point location is described. A small, active sonic "pill" for ultimately measuring deeper body temperatures is also discussed. Its sonic telemeter frequency is controlled by the temperature sought.

Data are presented which relate the orientation of the animal (dog), i.e., parallel or at right angles to a polarized rf field, to the rate and extent to which body temperature is elevated. Values relating exposure (intensity and duration) to survival are offered with the objective to derive L.D. 50 levels when more data accumulate. Smaller animals, such as guinea pigs, tolerate somewhat higher dosages of irradiation at 200 mc. Furthermore, orientation of the guinea pig with respect to the X or Y planes of propagation matters less in body temperature elevation than for a larger animal having a more extensive longitudinal axis.

Autopsy data suggest that the hollow viscera heat differentially high and the stomach and liver may be considered "hot spots." Damage to the gastro-intestinal wall is consistent with symptoms of digestive upsets observed in irradiated survivors.

Effects of 2450 mc Microwaves in Dogs, Rats, and Larvae of the Common Fruit Fly

Gordon W. Searle, Roger W. Dahlen,
Charles J. Imig, Charles C. Wunder,
John D. Thomson, John A. Thomas,
and William J. Moressi
Department of Physiology
College of Medicine
State University of Iowa
Iowa City, Iowa

This report deals with three lines of study of the biological effects of 2450-mc cw electromagnetic radiation. The first is concerned with irradiation of the heads of dogs as an elaboration of previously reported studies (1). The second takes up again the problem of the relative effects of microwaves upon the testes of rats in comparison with the effects of infrared heating. Refinements of techniques previously used in this laboratory (2) are being employed. The third is a preliminary report of studies on the growth of fruit fly larvae under microwave irradiation. Growth constants for this form have been estimated as in similar studies in our department where the influence of heat and gravitation upon growth have been determined (3, 4).

Head Irradiation in Dogs

These studies were started in dogs with the idea that the spaces filled with cerebrospinal fluid in and around the brain might be foci of heat accumulation as fluid-filled hollow viscera in the rabbit had been reported to be (5). Some tendency to support this hypothesis has been reported (1). Evidence for functional changes referable to higher centers have been sought. A group of 9 mongrel dogs anes-

187

thetized with sodium pentobarbital were irradiated with the maximum output of a 125 w generator (Raytheon Microtherm, Model CDM-10 with "C" director) which gave a calculated field density of 0.8 w/cm^2 at the surface of the exposed scalp in the center of the field. Systolic and diastolic blood pressure were recorded by a pressure transducer from the femoral artery. Cerebrospinal fluid pressure was similarly recorded from the cisterna magna. These values together with the heart rate and calculated mean arterial blood pressure are shown in Figure 1 plotted as functions of the per cent of the total survival time. The survival times for this group of animals ranged from 180 to 390 min (mean = 311 min). The heart rate increased steadily throughout the exposure period. The diastolic pressure was significantly higher than the initial level at the half and three-quarter marks, then fell. The systolic pressure did not change significantly until the terminal quarter of the period. No marked change occurred in the cerebrospinal fluid pressure throughout. The fatal termination appeared to be one of circulatory collapse. The rapid pulse and the falling pulse pressure could have been the result of peripheral vasodilation set in motion to maintain normal limits of body temperature. Panting, which was seen almost from the beginning of the exposures, was inadequate for controlling the body temperature. Panting may have contributed to the shocklike picture by contributing to water loss through evaporation.

Since the head of the dog is a complex mass with respect to

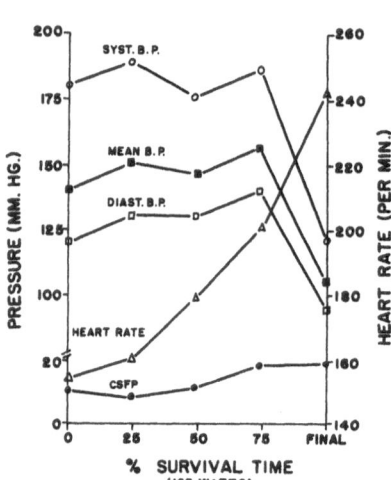

Fig. 1. Systolic pressure (○), mean arterial pressure (■), diastolic pressure (□), heart rate (△), and cerebrospinal fluid pressure (●) in dogs during irradiation of the head with 2450-mc cw microwaves.

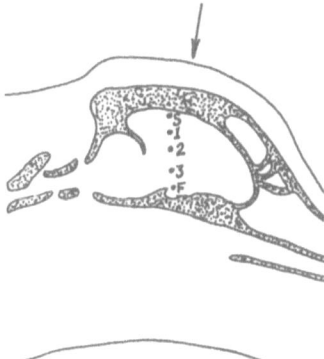

FIG. 2. Midsagittal section of the head of the dog to show positions of thermocouples at various depths in the brain.

reception and distribution of microwave energy, it was considered desirable to measure the temperature changes within the cranium with respect to depth, irrespective of the specific structure of the brain, and to note the time courses of temperature development also. Accordingly, thermocouples were placed as shown in Figure 2. The thermocouples were inserted through burr holes near the midline of the calvarium. The most superficial, "S," was placed just beneath the dura. Thermocouples designated "1," "2," and "3" were placed at equal intervals in order of depth between "S" and the deepest one, "F," on the floor of the cranial vault. Measurements were made from these and from a rectal thermocouple in 5 anesthetized dogs at the maximum output of the 125 w generator as before until death. The mean survival time for this group was 256 min with a range from 210 to 300 min. The results are shown in Figure 3 where the time axis is again expressed in terms of per cent of the total survival time. The most superficial thermocouple recorded the highest temperatures after the first part of the period. The others followed in order of depth with the rectal temperature a little below the lowest brain temperatures.

It was considered desirable to contrast such a pattern in the living, anesthetized animal with the pattern of temperatures in similarly irradiated dead animals. The experiment was, accordingly, repeated in 5 dogs immediately after sacrifice by sodium pentobarbital overdosage. The results are shown in Figure 4. All of these exposures were 2 hr long. In this time all but the deepest cranial thermocouples had exceeded the scale of the recording potentiometer, the upper limit of which was 70°C. The rectal temperature was

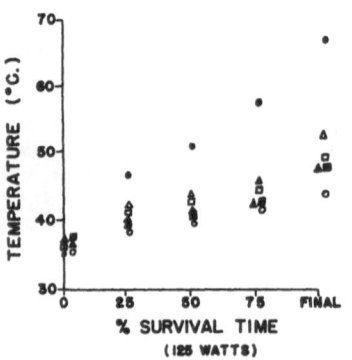

Fig. 3. Courses of temperatures in the various depths of the brain: Subdural (●); No. 1 (△); No. 2 (□); No. 3 (▲); and Floor (■); and in the rectum (○) of living dogs irradiated with the Raytheon Microtherm.

steady or slowly falling off throughout. This demonstrates, by a greater spread, the order of the decreasing heating rate from the superficial, near the radiation source, to the deep parts of the cranial contents. The importance of the convection distribution of the heat by the blood is emphasized in a negative way by the failure of the rectal temperature to rise in these dead animals.

Five dogs were irradiated in the same manner with a more powerful field (employing the Raytheon Model PGM-100 generator). In these experiments the central field intensity at the scalp was of the same order of magnitude as in the preceding (about 0.5 w/cm^2) but the output utilized (750 w) and the antenna horn made possible a more sustained power density out to the periphery of the scalp, hence a greater total amount of energy for the area exposed. In this case the anesthetized dogs died in the average time of 91 min (75 to 137 min). The pattern of temperature rise was much like that seen in

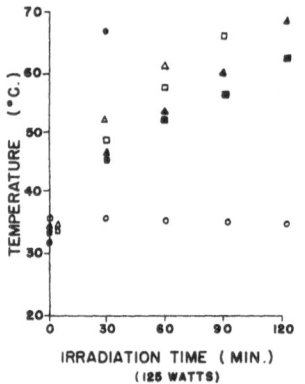

Fig. 4. Courses of temperatures in the various depths of the brain and in the rectum of dead dogs irradiated with the Raytheon Microtherm. (The symbols are identical to those in Fig. 3.)

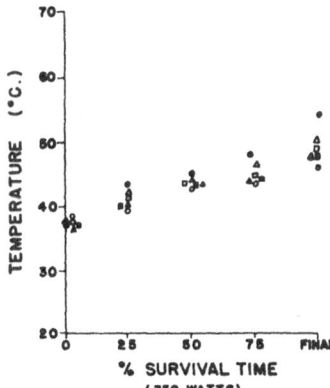

FIG. 5. Courses of temperatures in the various depths of the brain and in the rectum of living dogs irradiated with the Raytheon Model PGM-100 microwave generator. (The symbols are identical to those in Fig. 3.)

the experiments with the lower power when graphed against percentage of the survival time. The results are shown in Figure 5. The terminal temperatures, while only slightly higher than were obtained with the weaker generator, were achieved much more quickly.

Using this more powerful generator, the effects in dead animals were sought for further contrast between the more plainly physical effects and the physiologically modified effects in the living animal. The results are shown graphically in Figure 6 where the time course is carried to 120 min at which point all the temperatures in the head region exceeded the limit of the recording apparatus. The rectal temperature again remained approximately constant. This series of 5 dogs confirmed the fact that the rise in the rectal temperature of

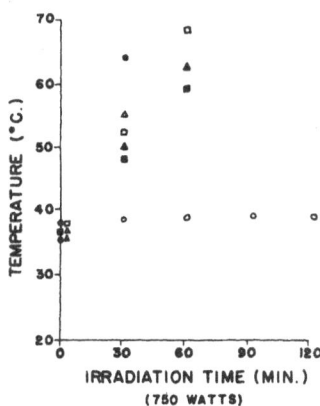

FIG. 6. Courses of temperatures in the various depths of the brain and in the rectum of dead dogs irradiated with the Raytheon Model PGM-100 microwave generator. (The symbols are identical to those in Fig. 3.)

the living animals was chiefly due to the convective function of the blood for heat distribution when the circulation is maintained.

Irradiation of the Testes in Rats

The testes of young adult, male, albino rats (Holtzman) were exposed to the output of a clinical microwave diathermy apparatus (Raytheon Microtherm, Model CDM-10 with "C" director). The rest of each animal was shielded by means of a copper screen. The exposure was made with the distal end of the testis directed toward the antenna at the distance of 5 cm. One testis was used for recording the temperature in the distal, middle, and proximal part of the organ approximately in the longitudinal axis by means of needle thermocouples. The other testis was used for follow-up studies of histological damage. The arrangement is shown in Figure 7.

A similar arrangement was used with an infrared lamp (Burdick "Zoalite," Z-12) as the source of radiation. In this case an asbestos shield was used to protect the rest of the animal from the radiation. The follow-up studies were the same.

The stages of histological damage to the testes were classified as follows:

STAGE 1: Diminished germinal elements in some seminiferous tubules, sloughing of germinal elements, and intertubular edema with enlargement of intertubular spaces.

STAGE 2: Some seminiferous tubules devoid of germinal elements. Germinal elements may take on a "coagulated" appearance.

STAGE 3: More tubules completely sterile.

STAGE 4: Nearly complete degeneration of most of the seminiferous tubules with only Sertoli cells and basement membrane remaining. Presence of multinucleated "giant cells" with dark staining nuclei.

One of these stages of damage was assigned to each testis examined on the basis of microscopic examination of several fields in slides of longitudinal sections stained with hematoxylin-eosin stain.

Each irradiation modality was varied in power output to maintain, as nearly as possible, constant temperature for a 15-min period. The central thermocouple was used as the guide in this. Thus, infrared and microwave exposures were equilibrated by the tempera-

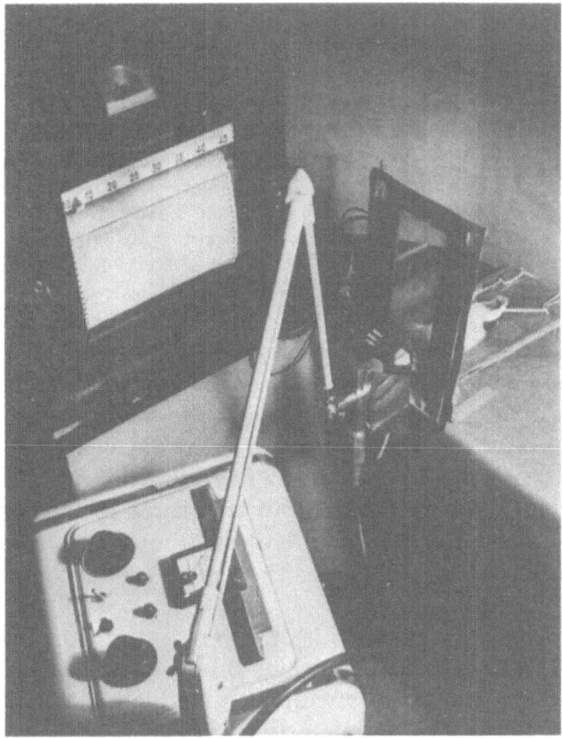

Fig. 7. Arrangement for irradiation of the testes of the rat: recording potentiometer for temperatures above; Raytheon Microtherm below; and animal positioned before the antenna-director at right. Three needle thermocouples are inserted into one testis of the animal.

ture effects. In the first series of rats the stages of histological damage were assessed one hour after the irradiation period. Two to five rats were used in these experiments and the results are shown in Figure 8. For comparison between the two types of radiation the results at different temperatures have been graphed as vertical columns showing the average stage of damage for the group receiving microwave irradiation and infrared, respectively. It can readily be seen that no damage was visible at this interval of time following exposure in the infrared heated groups. Damage from microwave irradiation was present in all but the lowest temperature used.

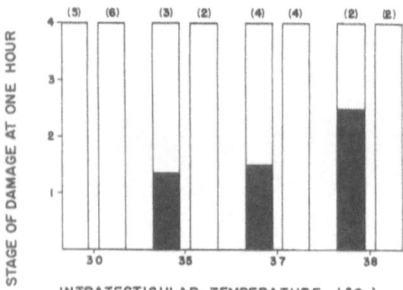

FIG. 8. The stages of histological damage in testes of rats sacrificed 1 hr after 15 min exposure at various intratesticular temperatures. The left-hand column of each pair represents the average stage of damage in the microwave exposure; the right-hand, the infrared. Numbers of animals in each group are indicated in parentheses above each column.

In Figure 9 the results in a second series of rats 2 days following similar 15-min exposures are shown. Damage in all groups but one was seen although the lowest temperature in the preceding series was not used in this one. The damage by microwave exposure was more extensive by histological criteria at 2 days than at 1 hr. The damage that was found at 2 days in the case of infrared irradiation was still, apparently, much less than that resulting from the microwaves. Experiments in progress will extend these observations to 4- and 6-day intervals following exposure.

Histological examination is not very satisfactory for grading damage to the interstitial cells which produce testosterone. Accordingly, a sensitive test for testosterone production utilizing the fact that androgenic substances stimulate the production of fructose by the accessory reproductive organs (6) has been used. The fructose concentration of the anterior prostate (sometimes known as the "coagu-

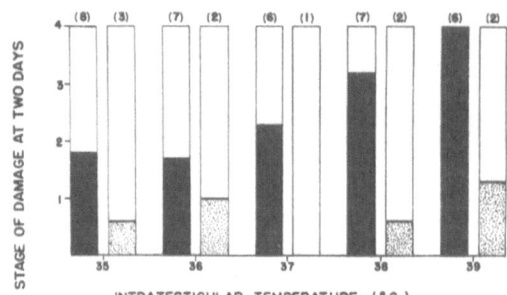

FIG. 9. The stages of histological damage in testes of rats sacrificed 2 days after 15 min exposure at various intratesticular temperatures. Same scheme as in Fig. 8.

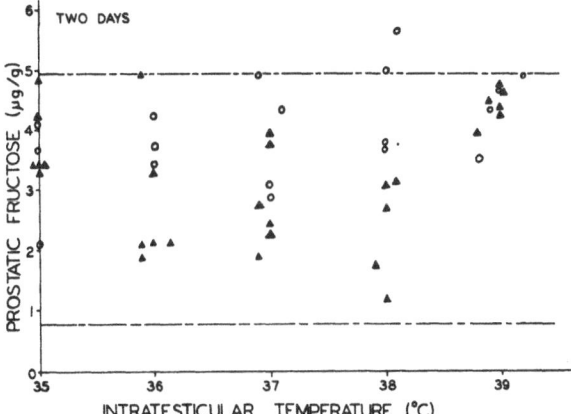

Fɪɢ. 10. Fructose concentration in anterior prostate of rats sacrificed 2 days after 15 min exposure of testes to microwaves (▲) or infrared (○) at various intratesticular temperatures. Broken lines indicate normal range in 15 rats.

lating gland" in the rat) was determined in untreated rats and in rats irradiated in the same manner as described above with microwave or infrared, respectively. The results at 2-days post-exposure are shown in Figure 10. At this point in time the values for fructose at all irradiation period temperatures fell within the normal limits (mean ± 2 standard deviations) derived from results on 15 animals. It was expected that there would be a drop in this value in the test animals, but a slight tendency for an increase was seen instead. This could have been due to a flushing out of testosterone from the interstitial cells as a result of the hyperemia produced. These observations are to be extended to longer periods following irradiation as in the case of the histological studies of the germinal elements. A fall in fructose to near zero is expected if testicular interstitial cell damage is marked.

Growth of Larvae of the Common Fruit Fly under Microwave Irradiation

The larval stage of the common fruit fly (Drosophila melanogaster, wild, red-eyed) is a convenient form for studying growth because of its relatively rapid and simple growth pattern. It has been used for the assessment of high gravitational effects as well as effects of

temperature upon growth in this laboratory (3, 4). The estimate of
growth of the larvae has been made by measuring the image or
shadow of individual larvae on a photographic plate (3). By assuming the larval shape to be an ellipsoid of revolution, the volume (V)
is calculated from length (l) and width or diameter (w) by the
formula

$$\pi w^2 l/6$$

The growth is exponential for about 3 days after hatching. On this
basis a growth constant (K) can be calculated from any two measurements of larval body volume over a time interval ($t_2 - t_1$) by the
formula

$$K = (\ln V_2 - \ln V_1)/(t_2 - t_1)$$

The larvae were grown in circular chambers containing a banana
agar as shown in Figure 11. This chamber was cooled by water
passing under, over, and around it. A thin latex membrane protected the culture from the water and allowed efficient cooling.

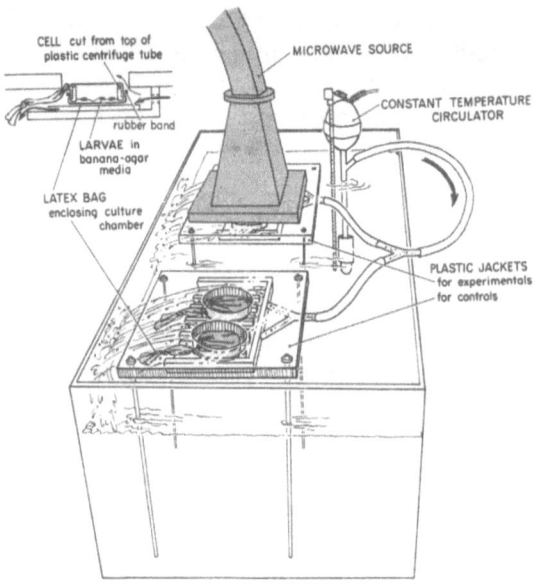

Fig. 11. Schematic drawing of apparatus for exposure of
growing fruit fly larvae to microwave irradiation at constant temperature.

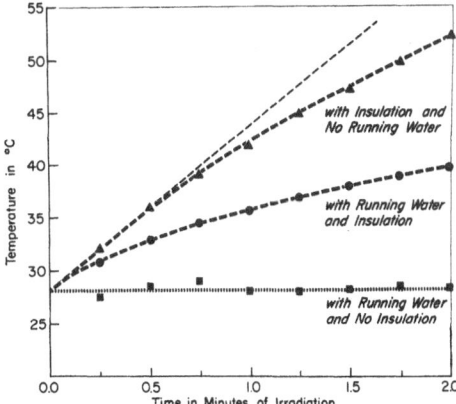

FIG. 12. Temperature courses in the banana agar medium within the growth chamber of the apparatus shown in Fig. 11 under different conditions. The lowermost curve represents the condition used in experiments.

Photographic images of the growing larvae were made at intervals by temporarily removing the larvae and placing them in a transparent chamber upon a photographic plate. The arrangement proved satisfactory for preventing a rise in temperature of the medium under continuous irradiation up to 68 hr. That energy was absorbed as heat without the cooling effect of the running water was proven by dummy runs graphically recorded in Figure 12. In these experiments, the chamber containing medium and covered only with a thin latex membrane was kept at 28°C by circulating water, but, when the water was kept from contact with the chamber by means of an asbestos insulation, the temperature rose steadily. In the latter case, the path of rising temperature rose at approximately the same rate for the first several seconds of irradiation as did the temperature in the insulated chamber with no water circulating around it. Thus, it seemed likely that the microwave energy was reaching the medium where the larvae were growing.

As shown in Figure 11, a control culture was grown in parallel during each experiment. When the volumes of control and irradiated groups were compared, the results shown in Figure 13 were obtained. Here it is seen that the volumes as indices of growth followed the same path as long as the temperature of the larvae was satisfactorily kept from rising. The irradiated animals showed a leveling off at a smaller volume than controls after about 68 hr. The reason for the deviation at this point may have been due to the fact that the larvae became more mobile at this age and were seen to climb onto the undersurface of the rubber membrane where it was more likely

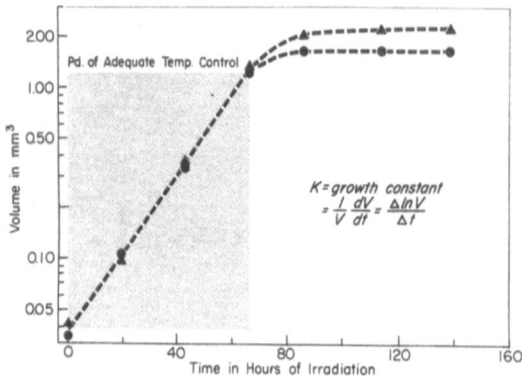

FIG. 13.   Courses of growth in control (▲) and irradiated
(●) fruit fly larvae.   The formula used for calculating
the growth constant, K, is also shown.

that they would be warmed above the 28°C which was the optimum
temperature for their growth.

In two experiments with field intensity of 0.3 w/cm² the ratio of
the growth constant of the radiated larvae to the growth constant of
the control group was 0.99 ± a standard error of 0.058.   In 8 experiments with field intensity of 1.0 w/cm² the same ratio was 1.02
± 0.025.   Without the provision for cooling, this intensity gives a
50% mortality in 10 min.   Thus growth of larvae was not significantly
affected by the field of microwave irradiation at 12.25 cm wavelength
when optimum temperature was maintained.   More intense fields are
to be studied with the same experimental design.

Summary

The courses of blood and cerebrospinal fluid pressures and heart
rate have been studied during the period of survival in dogs subjected
to 2450-mc cw microwave irradiation of the head.   Time and
depth gradients of heat development in the head also have been determined.   Comparisons between the gradients of heat development
in the head and in the rectum have been made between living and
dead dogs.   Studies on irradiation in rat testes have been made with
the average intratesticular temperature during the irradiation period
correlated with damage to the tissue.   The effect on the germinal
epithelium of microwave irradiation has been compared with the

effect of infrared heating. Anterior prostate fructose has also been used as an index of androgen production after both types of irradiation. The larvae of the common fruit fly have been irradiated in an apparatus which permits maintenance of constant temperature. Under such continuous exposure, growth rates over 3-day periods have not thus far been affected when optimum temperature for growth has been maintained.

References

1. Searle, G. W., Imig, C. J., and Dahlen, R. W., "Studies with 2450-mc CW Exposures to the Heads of Dogs," *Proc. Third Annual Tri-Service Conf. on Biol. Effects of Microwave Radiating Equipments*, RADC-TR-59-140, August 1959, p. 54.

2. Imig, C. J., Thomson, J. D., and Hines, H. M., "Testicular degeneration as a Result of Microwave Irradiation," *Proc. Soc. Exptl. Biol. Med., 69,* 382 (1948).

3. Wunder, C. C., "Gravitational Aspects of Growth as Demonstrated by Continual Centrifugation of the Common Fruit Fly Larvae," *Proc. Soc. Exptl. Biol. Med., 89,* 544 (1955).

4. Wunder, C. C., Herrin, W. F., and Crawford, C. R., "Combined Influence of Gravity and Temperature upon Growth of Fruit Fly Larvae," *Growth, 23,* 349 (1959).

5. Hines, H. M., and Randall, J. E., "Possible Industrial Hazards in the Use of Microwave Irradiation," *Elec. Eng., 71,* 879 (October 1952).

6. Mann, T., and Lutwak-Mann, C., "Secretory Function of Male Accessory Organs of Reproduction in Mammals," *Physiol. Rev., 31,* 27 (January 1951).

The Effect of 2450 mc Radiation on the Development of the Chick Embryo

CLAIRE VAN UMMERSEN
Tufts University
Medford, Massachusetts

INTRODUCTION

THE EFFECT of microwave radiation upon the crystalline lens of the eye has been investigated for several microwave frequencies, particularly 2450 mc. Daily et al. (1) reported the induction of lens opacities in the dog, and the same year Richardson et al. (2) reported similar results in the rabbit. The cataractogenic effect of this frequency has been confirmed for the dog eye by Salisbury et al. (3) and for the rabbit eye by Daily et al. (4), Williams et al. (5), and Carpenter et al. (6, 7).

The normal process of growth in the lens is similar in several respects to the process of embryonic development. The lens grows throughout life by proliferation of new cells and their subsequent differentiation into lens fibers, the two processes taking place concurrently. In the developing embryo, proliferation and differentiation are taking place at the same time and at a relatively rapid rate. It therefore became of interest to test the effect of microwave radiation on the developing embryo.

Materials and Methods

All experiments were carried out on chick embryos which had been incubated at 39°C to approximately the 48-hr stage of development. The microwave source was a Raytheon CMD4 Microtherm, which emits continuous wave radiation at a frequency of 2450 mc. A corner reflector antenna, the Microtherm Director C,

201

was mounted at the top of an anechoic chamber measuring 12 x 12 x 15 in. and lined by microwave absorbent. The egg was supported on a Lucite table 2 in. beneath the antenna.

Although the radio-frequency energy dissipated in the microwave absorbent during irradiation caused it to become heated, uniform incubator temperature was maintained by means of an exhaust fan mounted at the top of the chamber. Temperature was monitored by a thermistor-thermometer.

Before results could be assayed, it became necessary to know approximately how much radio-frequency energy was being received by the egg. Input power meter settings were therefore equated with power density measurements made calorimetrically in the anechoic chamber. An egg shell of known volume and profile area, filled with physiological saline and placed in the position of an egg during irradiation, served as the calorimeter. A hypodermic needle thermistor probe was inserted through the shell so that its tip approximated the normal position of the embryo. The energy absorbed by the embryo could then be determined for any selected power level. The maximal output of the generator yielded a power density of 400 mw/cm². Measurements at 70 and 50% of maximal output were proportionately less, 280 and 200 mw/cm², respectively. At these power levels, embryos at the 48-hr stage were irradiated through the intact shell, thus causing minimal interference with their natural internal millieu.

Temperature increases within the egg during irradiation were measured by inserting the thermistor probe through the blunt end of the egg with the tip of the needle placed in the yolk directly below the embryonic disc. After exposure, the thermistor probe was withdrawn, the needle hole sealed with Scotch Tape, and the egg returned to the incubator. These embryos developed in a manner similar to those exposed for identical periods of time but without penetration by a thermistor probe.

Thermal gradients within the egg were also determined. Nonfertile eggs heated in the incubator to 39°C were punctured in the minor axis and each hole sealed with Scotch Tape. The thermistor probe was inserted from below through the Scotch Tape and adjusted so that its tip was approximately 4 mm from the shell of the upper surface of the egg. The temperature at this point was recorded during the minimal lethal exposure period for each of the power densities used. After the power was turned off, temperatures were recorded after 1, 5, 10, and 20 min as the egg cooled. Similar measurements were made at positions 8, 16, and 28 mm below the shell.

After irradiation, the eggs were allowed to continue development for another 48 hr, at which time the embryos were removed and examined before being preserved and stained. To facilitate correlation of gross and microscopic observations, each embryo was photographed before being embedded in paraffin and sectioned serially at 10 micra.

Results

The total number of eggs incubated was 507, of which 366 were exposed to radiation, 109 served as controls, and 32 were not fertile. At the maximal power of 400 mw/cm², embryos which were irradiated for 1 to 4 min showed no appreciable deviation from the control embryos. Irradiation for 4½ to 5 min caused abnormalities and if prolonged for another half minute became lethal. At lower power levels, the difference between exposure periods which were ineffective and those which were lethal was similarly a matter of no more than 2 or 3 min.

The results of all experiments are summarized in Table I. Of the 109 control embryos, 106 continued their development normally, as did also 142 embryos which received ineffective doses of radiation. Of the 103 embryos which died, some had continued to develop for only a few hours after irradiation and others had continued for a day.

At 48 hr of development, the stage at which embryos were irradiated, the following features are evident: Torsion has occurred to a level slightly posterior to the heart so that the anterior portion of the embryo lies on its left side. Lateral undercutting of the body folds is proceeding so that the gut is being floored in and the embryo is being isolated from the yolk. The brain has just begun to differentiate from three into five vesicles. The primary optic vesicle has invaginated to form a cup and the lens is in vesicle form. Two visceral arches are present, with the beginning of a third. Although blood circulation has been established, the heart has not yet begun to be partitioned into right and left sides. It is essentially a tubular organ located at the level of the rhombencephalon. Lying outside the body cavity and bulging to the right of the embryo, it has begun to undergo the flexions and torsions which will bring its regions into their adult relationship. The region posterior to the heart is relatively undifferentiated. (See Figs. 1 and 2.)

TABLE I

Irradiation of Chick Embryos at 48 Hours of Incubation

Power (mw/cm²)	Time (Min.)	Number of Cases	Results			
				Abnormalities		
			Normal	General	Restricted	Dead
Controls		109	106			3
400	1	10	10			
400	2	8	8			
400	3	10	10			
400	4	22	20		2	
400	4½	46	9	26	4	7
400	5	32	2	17	4	9
400	5½	10				10
400	6	10				10
280	4	9	9			
280	5	15	15			
280	6	15	15			
280	6½	10	10			
280	7	17	4	8	2	3
280	7½	18		8	1	9
280	8	32		17	8	7
280	8½	10		5		5
280	9	20				20
200	8	10	10			
200	10	10	10			
200	12	9	9			
200	13	14	1	10	1	2
200	14	14		4	4	6
200	15	15				15
Totals		475	248	95	26	106

At 96 hr of incubation, the time at which all embryos were preserved for study, the normal chick embryo exhibits the following characteristics: It is C-shaped and lying on its left side; torsion, therefore, is complete. The body folds have isolated the embryo from the yolk except in the region of the yolk stalk. The brain has become compartmented into five secondary vesicles, and its walls have thickened. Hypophysis and epiphysis are well defined. The two layers of the optic cup are beginning to differentiate into pigment and neural layers. Lens fibers have elongated so that they nearly occlude the primary lens cavity. Four visceral arches are present. The heart has come to occupy a position more caudal with reference to other structures and lies at the level of the wing buds. The partitioning of the heart into four chambers has already begun, the interatrial and interventricular septa being present but not complete. The myocardium has thickened, particularly in the ventricular region. Lateral undercutting of the body folds has occurred and the heart is thus enclosed within the body. The anterior and posterior appendage buds appear as paddle-shaped outgrowths of the body wall directed ventrad and slightly caudad. The allantois is a well formed sac evaginated from the hind gut. (See Figs. 3 and 4.)

In the 121 embryos in which irradiation had interfered with development, the anomalies fell into two categories:

1. General inhibitory effects: In these embryos, gross abnormalities were observed which were not restricted to any particular region of the embryo. The most obvious effect was the smaller body size (Figs. 5 and 6), which resembled a 72-hr embryo rather than a 96-hr one. Torsion had progressed posteriorly only as far as the wing bud. The brain possessed only the same degree of differentiation present when the embryo was irradiated. The eyes were abnormally small and frequently were atypical in shape (Figs. 5, 6, and 7). Wing buds frequently were absent; when present, they were of subnormal size. The development of hind limb buds had been suppressed in almost every embryo. Only three visceral arches, instead of four, were present. The heart showed no evidence of partitioning into four chambers and its myocardium was extremely thin (Fig. 8). Posterior to the level of the heart, the degree of development equalled that of 48-hr embryos. Neither allantois nor tail bud had developed.

2. Posterior inhibitory effects: These embryos characteristically exhibited deficiencies only in the region of the body posterior to the wing bud, a region which at 48 hr of incubation has undergone little differentiation. Microwave radiation seemed to have inhibited the differentiation which would have taken place in the ensuing 48 hr. These embryos either had an anterior portion normal to the 96-hr stage but with the posterior part of the body lacking (Figs. 9 and 10) or there were anterior defects similar to those already described above and the posterior body region consisted of what appeared to be an undifferentiated thin-walled balloonlike vesicle (Figs. 6 and 11).

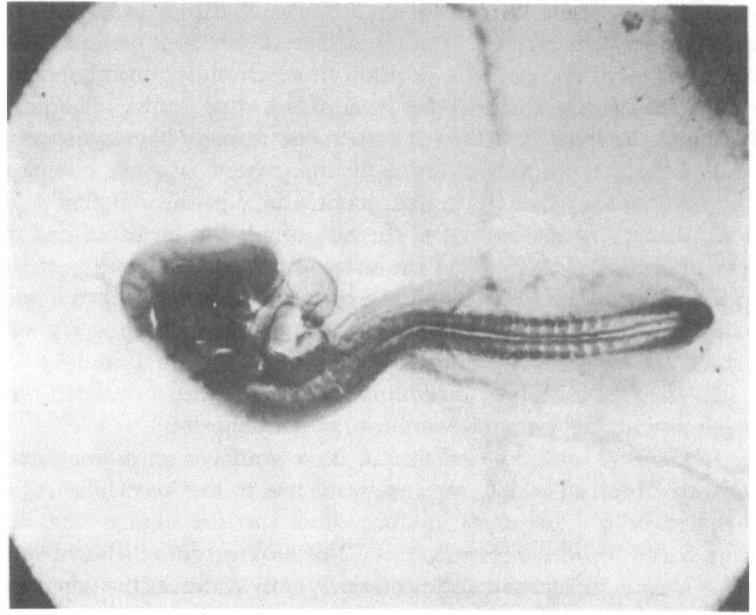

FIG. 1. Normal 48-hr chick embryo, whole mount. 12X.

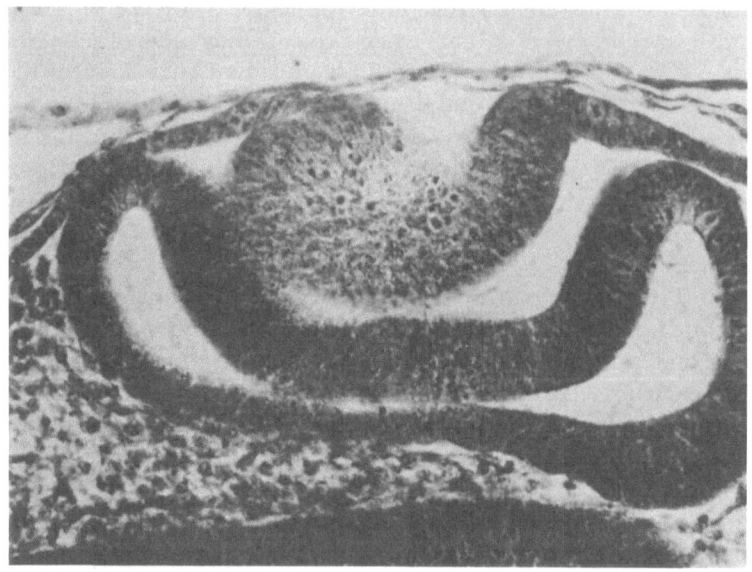

FIG. 2. Transverse section through the optic cup and lens
vesicle of a normal 48-hr embryo. 200X.

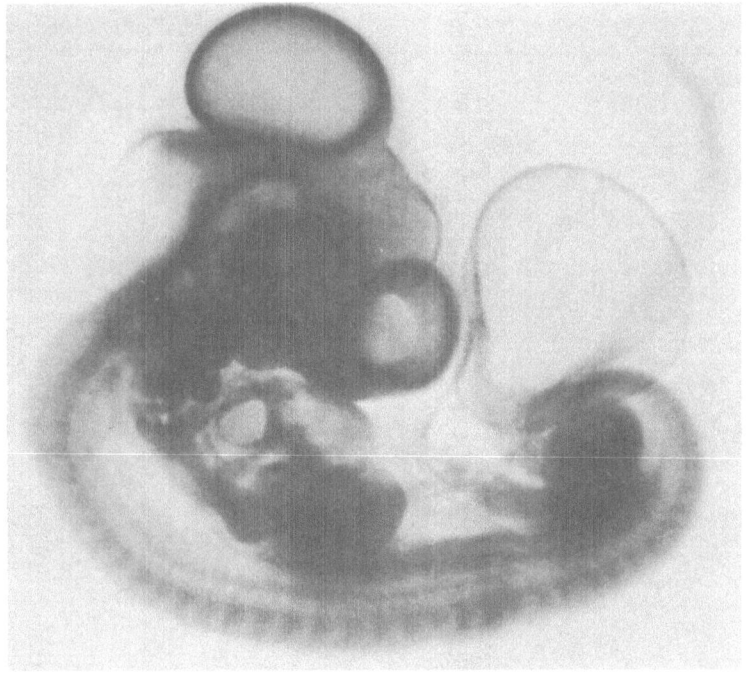

FIG. 3. Control 96-hr embryo, whole mount. 8X.

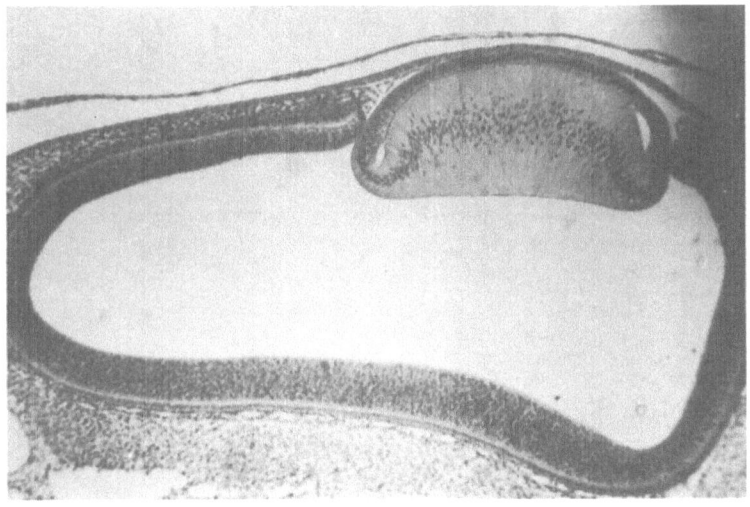

FIG. 4. Transverse section through the optic cup and lens of
a normal 96-hr embryo. 60X.

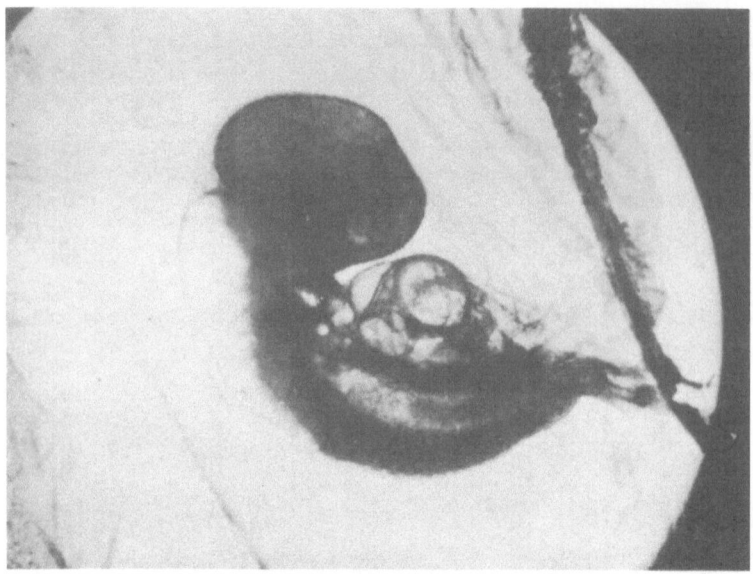

FIG. 5. 96-hr embryo which had been irradiated 7 min at
280 mw/cm², whole mount. 12X.

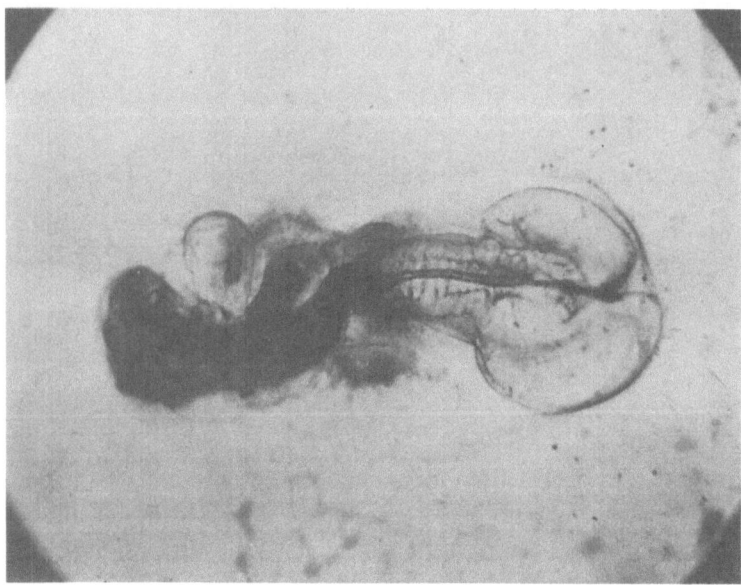

FIG. 6. 96-hr embryo which had been irradiated 8 min at
280 mw/cm², whole mount. 12X.

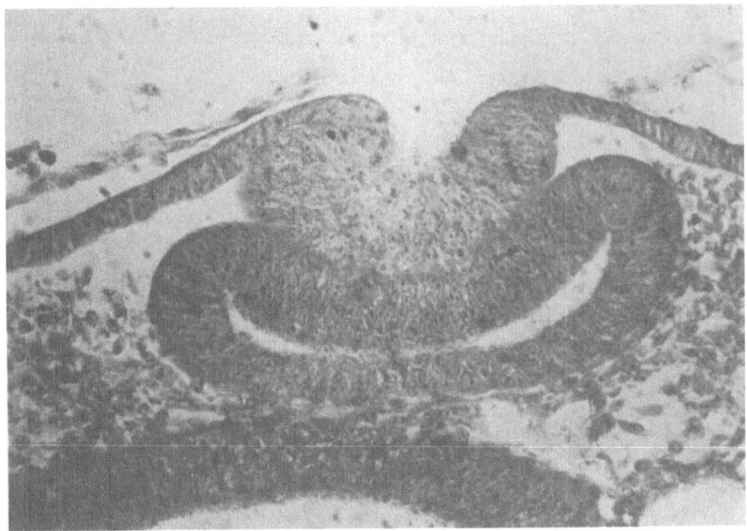

FIG. 7. Transverse section through the optic cup and lens vesicle of an embryo which had been irradiated 5 min at 400 mw/cm². 200X.

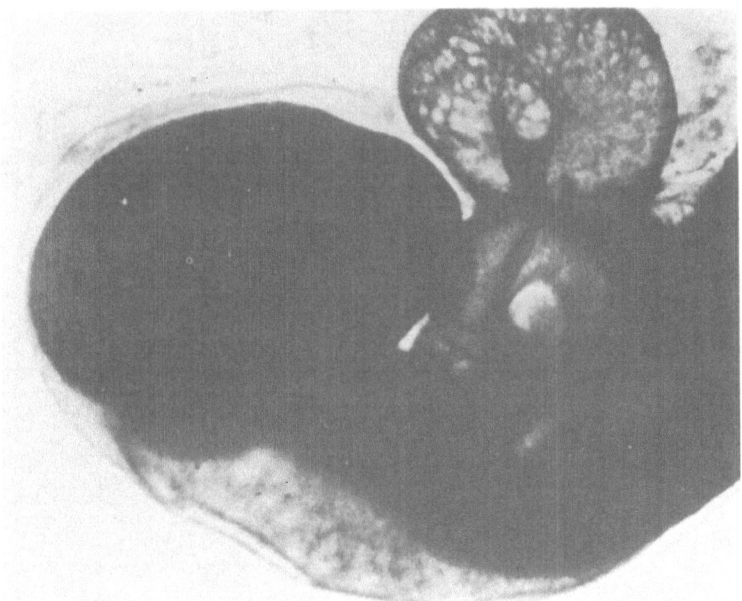

FIG. 8. Anterior region of a 96-hr embryo which had been irradiated 5 min at 400 mw/cm², whole mount. 12X.

Claire Van Ummersen

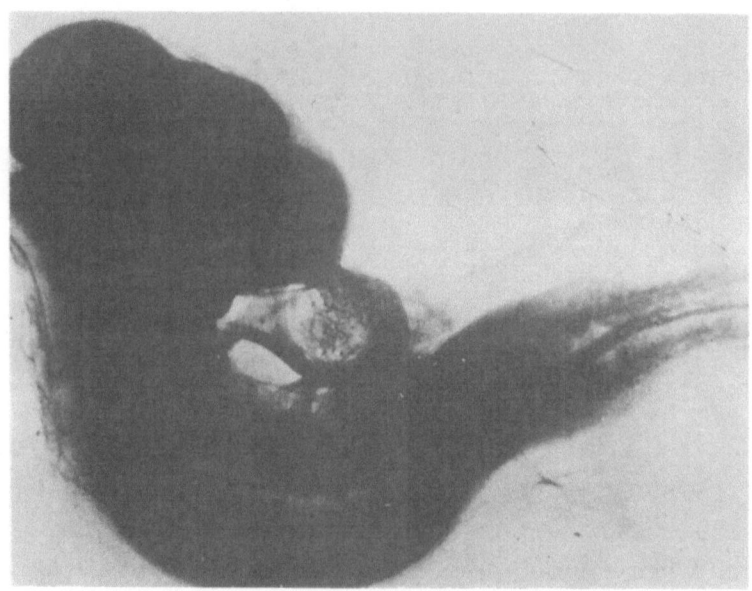

FIG. 9. 96-hr embryo which had been irradiated 14 min at 200
mw/cm², whole mount. 12X.

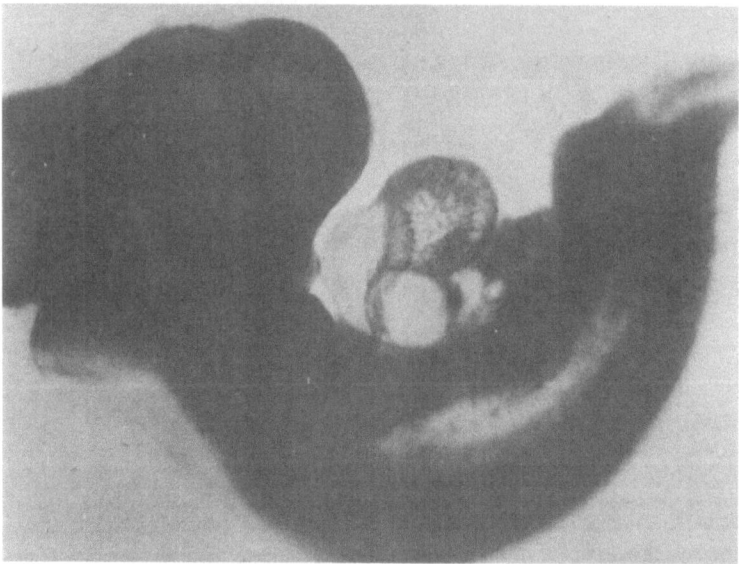

FIG. 10. 96-hr embryo which had been irradiated 13 min at
200 mw/cm², whole mount. 12X.

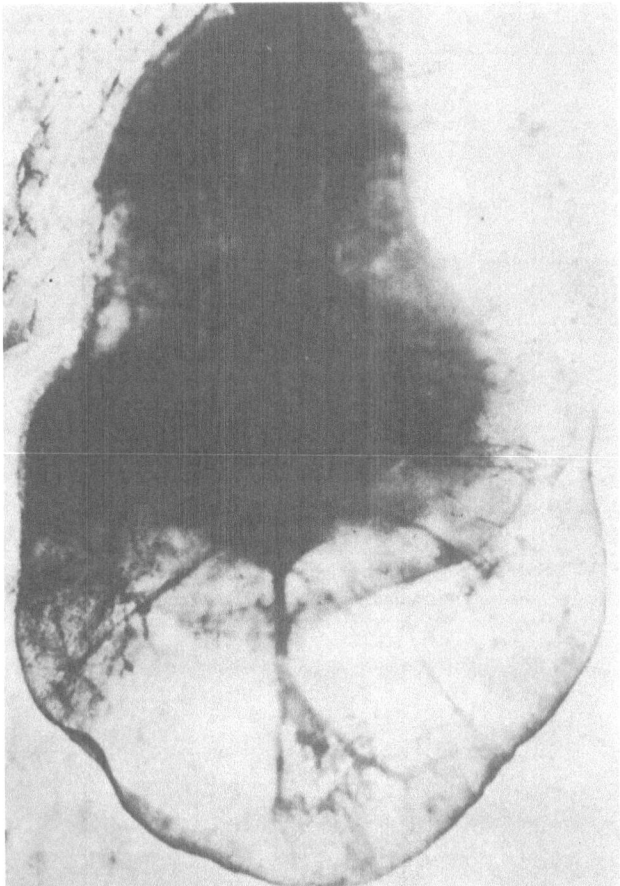

FIG. 11.   Posterior region of a 96-hr embryo which had been irradiated
5 min at 400 mw/cm², whole mount.   12X.

Microscopic examination of serial sections of the former group
showed that the anterior region developed normally but that posterior
to the wing bud some atrophy as well as inhibition of growth had
occurred.   The latter group of embryos exhibited brain, eye, and
heart abnormalities similar in nature to those already described but
posteriorly the effects were far more extensive.   In many embryos,
the neural tube posterior to the wing bud had developed as a multiple structure (compare Figs. 12 and 13 with Figs. 14 and 15) or had
failed to close.   The wing bud, if present, was directed laterad instead of ventro-caudad and there were no posterior limb buds.   The

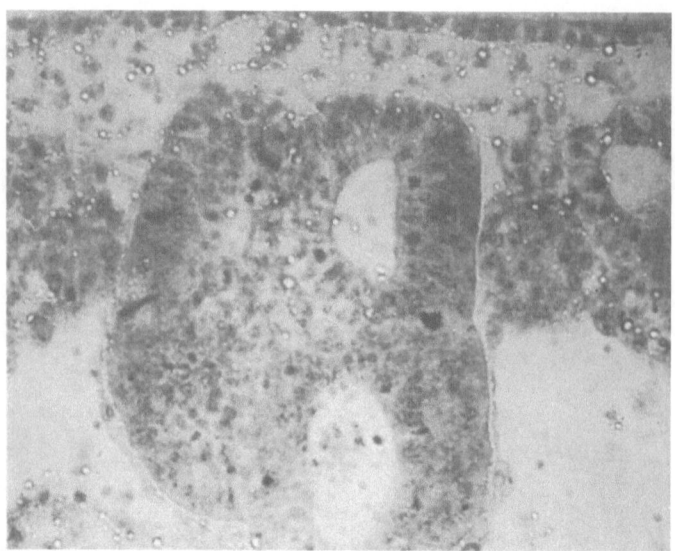

FIG. 12. Transverse section through the neural tube of an embryo which had been irradiated 5 min at 400 mw/cm². 200X.

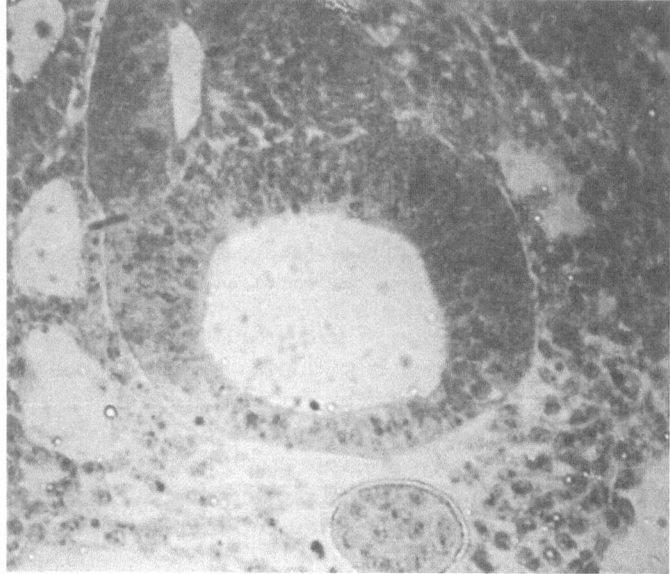

FIG. 13. Transverse section through the neural tube of an embryo which had been irradiated 8 min at 280 mw/cm². 200X.

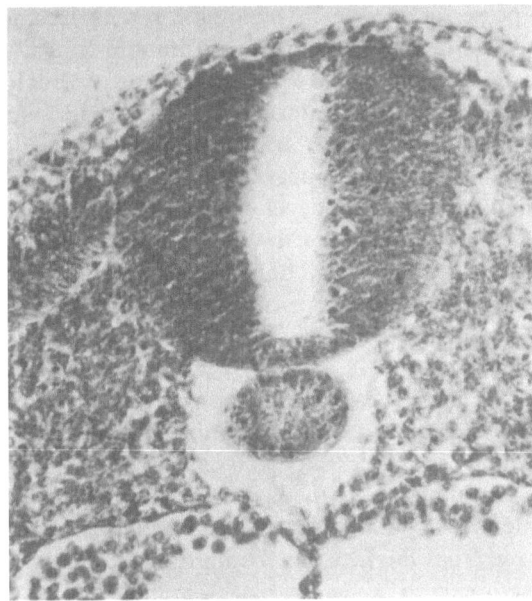

FIG. 14. Transverse section through the neural tube of a normal 48-hr embryo. 200X.

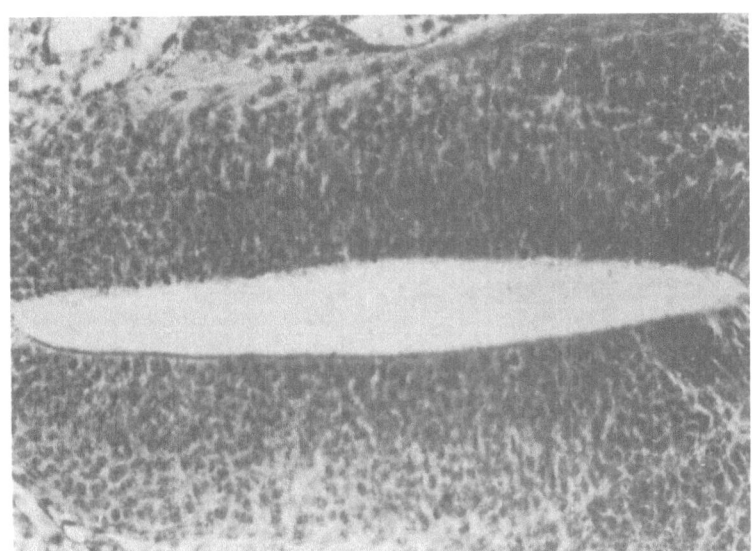

FIG. 15. Transverse section through the neural tube of a 96-hr control embryo. 200X.

balloonlike posterior vesicle in histologic section seemed to be of double nature, one part on each side of the mid-line of the embryo. It appeared to comprise a large coelomic space which had arisen chiefly through failure of the lateral body folds to develop sufficiently to separate an embryonic from an extra-embryonic coelom.

A number of temperature measurements were made to determine the rise in temperature at the level of the embryo during irradiation. The period of time required for the egg to cool to incubator temperature was also recorded. The data are presented in Fig. 16. Regardless of the applied power or duration of exposure, a lethal dose coincided with a temperature of approximately 59°C in the yolk just below the embryo.

The thermal gradient across the egg was also determined. When placed in an incubator, an egg heats uniformly from all points on its surface inward to the center, but when exposed to the radio-frequency energy from an antenna, the media nearest to the antenna heat most rapidly, while the media farthest from the antenna heat most slowly. This is inferred from the fact that 1 min after termination of irradiation, a gradient existed across the egg. Twenty minutes later the entire egg was close to a uniform temperature (Table II).

Discussion

The experiments here reported demonstrate that the normal course of development in the early chick embryo is adversely affected

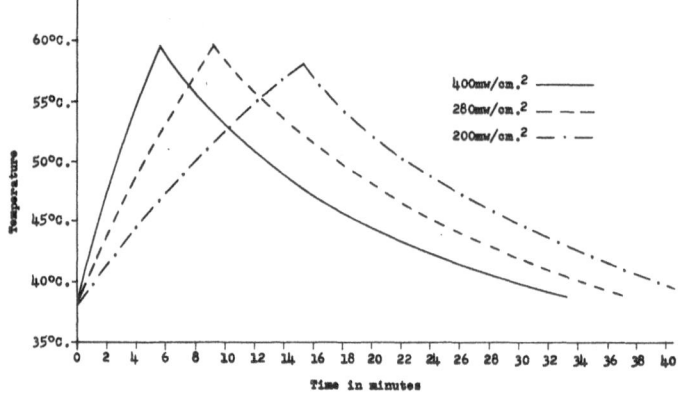

FIG. 16. Temperatures of the yolk below the embryo.

TABLE II

Thermal Gradients Within the Yolk

Distance from shell surface	Temperature during cooling period at:				
	0 min	1 min	5 min	10 min	20 min
A. Previous irradiation at 400 mw/cm² for 5½ min					
4 mm	58.7	56.6	53.3	46.0	40.6
8 mm		56.3	51.6	45.9	40.6
16 mm		55.4	47.7	43.7	39.9
28 mm		43.9	39.5	38.0	36.8
B. Previous irradiation at 280 mw/cm² for 9 min					
4 mm	59.3	57.8	53.1	45.5	40.7
8 mm		57.4	53.0	45.5	40.7
16 mm		56.7	52.6	44.9	40.7
28 mm		45.2	42.1	40.1	37.4
C. Previous irradiation at 200 mw/cm² for 15 min					
4 mm	57.8	56.2	51.5	44.0	39.6
8 mm		55.8	51.2	44.0	39.6
16 mm		54.0	50.8	43.8	39.5
28 mm		45.0	42.3	39.7	37.1

by relatively brief exposure of the embryo to microwave radiation at a frequency of 2450 mc. The only other experiments of this nature which have been reported have been those of Osborne (8, 9), who found that at no stage of development was the chick embryo affected by long exposures to 200 mc radiation. It should be noted, however, that at 200 mc the wavelength is approximately 5 ft. It has been pointed out by Vogelman (10) that if an object in the path of microwave radiation is to absorb any of this energy, its dimension should be at least one-tenth the wavelength. Inasmuch as the greatest dimension of an egg is considerably less than the requisite 6 in., it is quite probable that in Osborne's experiments the radio-frequency energy passed through the egg with none, or very little, of it being absorbed. It may be significant that Osborne found the temperature rise within the egg to be never more than 1°C.

At the 2450 mc frequency, approximately two-thirds of the radiation which enters the egg is absorbed within a depth of 1 cm. This distance coincides quite closely with the position of the embryonic area in the egg. At this frequency, there is also a significant increase in temperature which is related to the power density and the duration of exposure. All embryos in which effects were produced reached a temperature of approximately 55°C. The question arises as to whether the abnormalities are induced wholly as a thermal effect or whether they are due wholly or in part to some other factor.

Pincus and Fischer (11) showed that cultures of chick osteoblasts were killed in 3.5 to 6 min at temperatures exceeding 50°C, but it must be noted that these were isolated cells in tissue culture. On the other hand, the experiments of Szarski (12, 13) demonstrated that so long as the blood circulation was intact, all tissues of the embryo remained viable. After heating eggs in a high temperature incubator (45.5°C) for an interval of 1 to 16 hr, Szarski opened the eggs and cultured pieces from various organs of the embryo on nutrient media. All tissues removed from embryos in which the circulatory system was still functioning retained full viability and continued to develop when cultured. Heart and intestinal tissues remained viable 1 to 2 hr after heart stoppage. The cerebrum was slightly more sensitive, remaining viable for only 1 hr after cardiac failure. Not until 13 hr of high temperature incubation, or some 11 hr after cessation of blood circulation, did the extra-embryonic membranes begin to lose their viability. These results led Szarski to conclude that the loss of viability of the tissues was not caused by the heat per se, but resulted from the accumulation of products of metabolism and lack of oxygen following circulatory failure. Organs with a higher metabolic rate, therefore, succumbed sooner.

Except in a few instances, the present experiments corroborate Szarski's conclusion with regard to the dependence of tissue viability on intact blood circulation. In most cases, the tissues continued to grow by proliferation but usually did not continue to differentiate. In some embryos, however, the posterior region began to atrophy even though the heart continued to beat.

Irradiation of the rabbit eye, either with X-ray or with microwave, results in the formation of characteristic lens opacities. In like manner, irradiation of the chick embryo by X-ray or by microwave radiation causes the development of characteristic abnormalities. Developmental abnormalities in embryos exposed to X-ray as reported by both Schneller (14) and Reyss-Brion (15) were almost identical with those which are caused by microwave radiation. Both types of radiation act to retard or inhibit development. In both instances similar defects occur in the anterior region of the embryo. Differentiation of the brain is arrested, the optic cup invaginates only slightly and no lens fibers develop. Development of the allantois is inhibited by both types of radiation. Reyss-Brion (15) also described posterior neural effects identical to those described in this paper, i.e., open neural folds, or two or three rudimentary neural tubes grouped together. There is no mention in the literature of anything similar to the balloonlike tails described here.

No measurable heat is produced by X-ray and yet the abnormalities which are produced by microwave are identical to those induced by X-ray. This similarity of results suggests the possibility of a nonthermal influence exerted by microwave radiation which could be acting either concomitantly or synergistically with the obvious thermal effect of the radiation.

At 48 hr of development, the brain, the heart, the optic cup, and the lens are undergoing differentiation as well as growth by proliferation. It appears that irradiation at this stage has the effect of arresting further differentiation without at the same time, in most instances, affecting proliferation. The optic cup and lens vesicle were exceptions in a small number of embryos in which they neither proliferated nor differentiated any further.

There are present at the 48-hr stage the seventeenth to nineteenth somites, which contribute the mesoderm of the wing bud. In some embryos the wing bud developed but was atypically small. Structures such as the posterior limb buds, tail, and allantois, which normally develop in the region posterior to the wing bud, exist at the 48-hr stage only as potential areas. Likewise, the posteriormost region of the embryo consists of merely three germ layers, although neural folds are present and there are differentiated extra-embryonic blood vessels. The effect of microwave radiation on this region is to prevent differentiation of new structures from occurring and they therefore do not develop at all. Obviously a structure must become differentiated before it can grow. In the case of neural folds and extra-embryonic blood vessels, some differentiation has already occurred. Further differentiation is inhibited but, as in the case of the brain, heart, etc., these structures continue to grow by proliferation.

In summary, microwave radiation appears to inhibit cellular differentiation in the developing chick embryo. In structures which have already begun to differentiate, cellular proliferation continues, but no further degree of differentiation occurs. Structures which have not yet begun to differentiate fail to do so subsequent to irradiation, and their development is therefore suppressed.

References

1. Daily, L., Jr., Wakim, K. G., Herrick, J. F., and Parkhill, E. M., "The Effects of Microwave Diathermy on the Eye," *Am. J. Physiol.*, 155, 432 (Dec. 1948).

2. Richardson, A. W., Duane, T. D., and Hines, H. M., "Experimental Lenticular Opacities Produced by Microwave Irradiation," *Arch. Phys. Med.*, 29, 765 (Dec. 1948).

3. Salisbury, W. W., Clark, J. W., and Hines, H. M., "Exposure to Microwaves," *Electronics*, 22, 66 (May 1949).

4. Daily, L., Jr., Wakim, K. G., Herrick, J. F., Parkhill, E. M., and Benedict, W. L., "The Effects of Microwave Diathermy on the Eye of the Rabbit," *Am. J. Ophthalmol.*, 35, 1001 (July 1952).

5. Williams, D. B., Monahan, J. P., Nicholson, W. J., and Aldrich, J. J., "Biologic Effects of Microwave Radiation: Time and Power Thresholds for the Production of Lens Opacities by 12.3 cm Microwaves." USAF School of Aviation Medicine Report No. 55–94, Aug. 1955.

6. Carpenter, R. L., "Experimental Radiation Cataracts Induced by Microwave Radiation," *Proc. Second Tri-Service Conference on Biological Effects of Microwave Energy*, Rome Air Dev. Center, Air Res. and Dev. Command, Rome, N. Y. ASTIA Doc. No. AD-131-477, July 1958, p. 146.

7. Carpenter, R. L., Biddle, D. K., and Van Ummersen, C. A., "Progress Report," *Investigators' Conf. on Biol. Effects of Electronic Radiating Equipments*, Rome Air Dev. Center, Air Res. and Dev. Command, Rome, N. Y. ASTIA Doc. No. AD-214693, Jan. 1959, p. 12.

8. Osborne, C., "Studies on the Biological Effects of 200 mc," *Proc. Second Tri-Service Conf. on Biol. Effects of Microwave Energy*, Rome Air Dev. Center, Air Res. and Dev. Command, Rome, N. Y. ASTIA Doc. No. AD-131-477, July 1958, p. 196.

9. Osborne, C., "Studies on the Biological Effects of 200 mc," *Investigators' Conf. on Biol. Effects of Electronic Radiating Equipments*, Rome Air Dev. Center, Air Res. and Dev. Command, Rome, N. Y. ASTIA Doc. No. AD-214693, Jan. 1959, p. 20.

10. Vogelman, J., "Physical Characteristics of Microwaves as Related to Biological Effects," *Proc. Second Tri-Service Conf. on Biol. Effects of Microwave Energy*, Rome Air Dev. Center, Air Res. and Dev. Command, Rome, N. Y. ASTIA Doc. No. AD-131-477, July 1958, p. 9.

11. Pincus, G., and Fischer, A., "The Growth and Death of Tissue Cultures Exposed to Supranormal Temperature," *J. Exptl. Med.*, *54*, 323 (Sept. 1931).

12. Szarski, H., "Heat Resistance of the Chicken Embryo Tissue in vitro," *Bull. intern. acad. polon. sci., Classe sci. math. nat. Sér. B; Sci. Nat. (2) Zool.*, p. 45 (Nov.–Dec. 1939).

13. Szarski, H., "Sur la mort thermique de l'embryon de poulet," *Bull. intern. acad. polon. sci., Classe sci. math. nat. Sér. B; Sci. Nat. (2) Zool.*, p. 133 (Nov.–Dec. 1947).

14. Schneller, Sister M. B., "The Mode of Action of Hard X-Rays on the 33 and 60 Hour Chick Embryo," *J. Morphol.*, *89*, 367 (Sept. 1951).

15. Reyss-Brion, M., "La sensibilite differentielle de certaines ebauches de l'embryon de poulet aux rayonx X, a differents stades du developpement," *Arch. d'anatomie microsc. morphol. exptl.*, *45*, 342 (Oct.–Dec. 1956).

Specific Thermal Effects
of High Frequency Fields

Victor T. Tomberg
Biophysical Research Laboratory
Elmhurst, New York

THE MEDICAL PROFESSION has used high frequency fields in the short and microwave range for more than 30 years. Based on observations (1) by Shereshevsky and Whitney in the United States, Esau and Pätzold in Germany, and Stieböck and Tomberg in Austria, the treatment by such fields produces deep heating in tissues and biological material. This acts in a stimulative manner on the body functions and its healing tendency. This heating or Joule effect is due to the absorption of energy by matter in a high frequency field (2). There are several practical ways to place an object in such a field. It may be a condenser field, the interior of a coil and its inductive field, or the radiation field of a dipole or a hf-cavity. In all these applications, medically called diathermy, the quantity of the produced heat or temperature rise in any part of a treated object depends on a number of factors such as applied energy, loss factor of the object determined by its electrical conductivity and dielectric constant, thermal constants, texture, absorption behavior, and specific electrical phenomena of molecular vibration.

The applied energy is not only related to the effective watts per square centimeter of the radiating field, or watts per cubic centimeter in a bipolar field, but it is also dependent on the configuration of the field transmitting parts, frequency, wave form, and other high frequency characteristics. This can be determined by measuring the temperature rise under given conditions in phantoms using thermoelectric probes and other thermosensitive means.

It was found that the biological effect of short and microwave fields depends on the product: (applied energy) $\times$ (time of action), a product called dose. When the dose is small, one can expect a biologically stimulating or healing effect. However, when the dose is high, relative to a certain threshold which depends on the nature of the interfering object, a harmful or destructive effect can result. The former action of the field is often called "bio-positive," the latter "bio-negative."

Several investigators have claimed that aside from the bio-positive thermal action there seems to be an "electrical" biological action of nonthermal nature, because certain treatments and biological effects showed results at very low field intensities and without apparent temperature rise. In a number of these cases, cooling was also applied from the outside to eliminate any temperature rise within the object. The theory that some wavelengths—similar to light waves— behave bio-positively and others bio-negatively has also been advanced (1).

However, it has been found generally and through our special investigations in this field that any biological action of short and microwaves particularly at low energies can be traced to particular actions of the thermal Joule effect which gives rise to so-called "specific-thermal" effects.

These specific-thermal effects are a class of effects which exist besides ordinary thermal effects and electric nonthermal effects. They look like electrical nonthermal effects because they cannot be duplicated entirely by ordinary heat, e.g., in a water control-bath, and they exist even under cooling conditions when applied experimentally. They may appear under field intensity conditions where safety level measures by ordinary instrumentation and check methods indicate safe values of about 10 to 30 mw/cm^2.

We would like to describe here in short their origin and particularities (4, 5, 7–11).

The rise of temperature in an object placed in a high frequency electric field, e.g., between condenser electrodes, depends on the frequency, the electric conductivity, and the dielectric constant when the object is homogeneous, e.g., an electrolyte. It can be easily derived that there is a maximum energy absorption when $k = \epsilon\nu/2$, where k is the electric conductivity in electrostatic cgs, ν the frequency, and ϵ the dielectric constant. Biological liquids having an ϵ of 80 and a k of about 4×10^9 therefore show a maximum absorption and temperature rise for a frequency of 100 mc which is experimentally verified (Fig. 1). It should be noted that ϵ and k are in most cases temperature dependent and not real constants. Because biological objects, like cells, tissues, blood, and microbes, have a heterogeneous texture where the constituents have different ϵ and k, their respective temperature rise must be different for a certain applied wavelength. This means that in a mixture, or in a particle suspension and the like, temperature gradients show up when subjected to an electrical high frequency field. For evaluating their biological importance we had to find out if these temperature gradients are equalized and impeded by thermal convection and conduction within the biological object. Experiments show that they are not, because the rate of the heating effect is physically faster than the rate of the temperature equalization process, especially in objects of low thermal conductivity. Hence, cooling applied from the outside to an irradiated object cannot prevent the build-up of temperature gradients and thermal spikes in discrete areas and between particles or layers of different composition. This frequency-dependent specific action of the high frequency field on heterogeneous matter, which

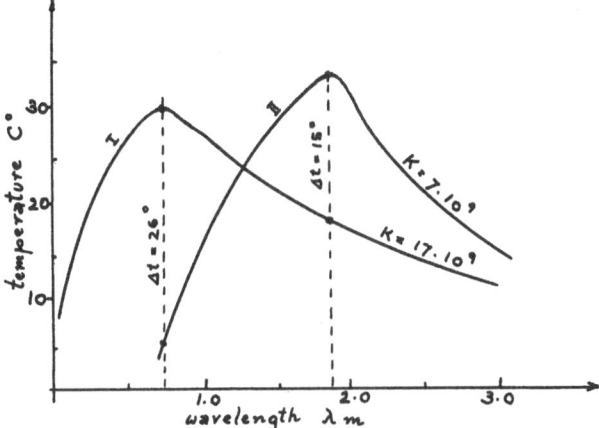

FIG. 1. The heating effect in two solutions. Solution I has an electrical conductivity K of 17×10^9 cgs, solution II of 7×10^9 cgs. At the wavelength of 0.70 m solution I is heating 26°C higher than solution II; however, at the wavelength of 1.85 m solution II is heating 15°C higher than solution I. Two vessels filled with each of the solutions and simultaneously irradiated would show a frequency-dependent temperature gradient according to their temperature-frequency specific-thermal absorption.

produces high temperature points in adjacent areas or between adjacent particles, is biologically the most important effect of bio-negative nature. This specific-thermal effect is even more developed in pulsed or modulated high frequency fields and a number of optimizing factors exist concerning their destructive nature. One of these factors is the polarization or resonating effect of molecules (Debye effect) at certain critical frequencies, and its connection with the electrical behavior of interboundary layers.

The temperature differences in discrete areas or between particles in an emulsion are difficult to measure. We used tiny thermocouples and micromanipulating procedures by placing samples under a microscope. We also used thermosensitive dyes introduced in the object to be examined and kept it at a predetermined temperature. We then observed that the highest temperature gradients arise in interboundary space and layers.

Actually, the increased use of high power transmitters poses interesting hazard problems due to the bio-negative action of short and microwave fields. Three kinds of actions and effects should be considered:

1. Ordinary thermal effects in more homogeneous material and biological objects.

2. Specific-thermal effects in heterogeneous biological material of low thermal conductivity due to:

(a) Irregular absorption which can be inherent to the heterogeneous structure of the exposed object or subject, or to irregular field intensity distribution with peaks near metallic objects, or dielectrics having resonating qualities which build up fields of high voltage gradients or standing waves.

(b) Temperature gradients with high peaks in some discrete areas of the exposed object when differences in electric conductivity and dielectric constant in the adjacent area are present. Low thermoconductivity helps to build up these phenomena, especially in interboundary space.

3. Electrical effects:

(a) Orientation effects, like the pearl-chain effect, which occur in emulsions when the two components, which may be solid in liquid or liquid in liquid, show different electrical characteristics. One of the components, particles or liquid droplets has the tendency to line up when free to move and not hindered by Brownian motion (3, 9).

(b) Frequency-dependent voltage and dielectric phenomena which are connected to the known physical and chemical effects of *Wien, Falkenhagen, Debye, Errera, Sack,* involve molecular resonance, anomalous dispersion and relaxation. Generally, they occur at frequencies over 300 mc, from a threshold frequency or in a very narrow frequency range.

(c) Chemical effects. In the range where significant quantum effects are possible (over 1000 mc), ionization effects with the formation of free radicals—O_2 and OH—may show up at high power levels, especially when high peak pulses are applied. The likelihood that they may occur is actually under study.

We said before that the biological effect is proportional to or dependent on the applied dose which is the power-time product. However, this is not true when the application time is very short or when high peak pulses and certain wave formations are applied. In all these cases, the factor power and the factor time have an exponent which may be higher or smaller than unity. This has been shown not only in high frequency experiments but also in the fields of ultrasonics, photobiology, and ionizing radiation. In some cases the biological effect appears to be dependent on the product $(P_2 - P_1)^n$ multiplied by t^m, where P_1 is a threshold value, P_2 the applied power in watts per square centimeter, and t the time of application.

What is the order of magnitude of the specific-thermal actions produced? When an isolated solid particle of diameter d and a thermal conductivity of ρ is irradiated while surrounded by an infinitely extended liquid like water, the maximum temperature difference between the particle and the liquid is equal to $Pd^2/4\rho$. Experimentally, this appears accurate as long as the diameter d is not smaller than about 80 μ. For particles smaller than this the maximum temperature difference is much higher, probably because the thermal conductivity and heat transfer condition are much smaller than the macroscopic value. Interboundary physical phenomena also play a more important part when the particle is smaller. Under practical conditions, we never deal with one particle alone in suspensions. In most cases, e.g., in microbial suspension, there are particles irregularly distributed within the liquid. In this case, we verified that the diameter d approaches the value of the colony or agglomeration of particles. At 10 mw/cm² power a difference of 10°C and more is possible when the particles are bigger than 80 μ. With heterogeneous biological material, such as cells or tissue, the same value may be reached under extreme conditions.

In biological or physical experiments, in order to discriminate between specific-thermal effects and those of an electrical nature, we perform the experiments at different predetermined temperature (as bias) and at different frequencies and intensities. Up to now, by the use of frequencies, pulses, and power conditions similar to those used in radar, we may say that the electrical nonthermal effects mentioned in class 3 physically exist, but that their biological importance, compared with the specific-thermal or even ordinary Joule effect, is negligible. This would not be the case if there were higher power, higher frequencies, or a very narrow frequency range which would be responsible for the molecular resonance phenomena.

In summary we would like to mention some biological findings due to the specific-thermal action of high frequency fields:

Permeability between cells increases more by irradiation than by applying ordinary heat and temperature increase.

Microbes or plant seeds can be stimulated in growth when the average temperature rise is negligible measured as a total thermal effect, although the individual cell would have a significant temperature rise within stimulating limits.

Microbes and plant seeds can be inhibited in growth or killed at temperatures which under ordinary heating conditions represent optimum growth conditions.

There are many more examples which are of more interest to biochemists.

From the standpoint of biological hazard, the specific-thermal effects of high frequency fields, especially under pulsed conditions, are more dangerous than the ordinary Joule effects and the electrical effects. We agree that the proposed safety level of 10 mw/cm^2 may be considered as a safe base as long as this average field value stays constant under free space conditions. Future situations in which intensification of the field will take place are not very remote, but these problems are difficult to anticipate. First of all, when a person walks into a regularly distributed field, the field is disturbed; metal parts and perspiration on the person may increase the voltage gradient on the skin level (Fig. 2). Other intensification factors include multiple reflections, standing waves, and resonant phenomena. It is known that the intensification factor can exceed 50; a number

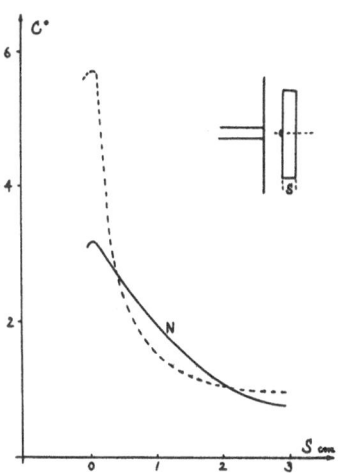

FIG. 2. A phantom specimen of S cm width is irradiated in a dipole field, first dry and then with a saline water drop at its center near the dipole. The temperature rise curves are different. In the first case it is curve N, and in the second case it is the dotted line curve. This illustrates how small amounts of water or humidity on a dry surface induce an irregularity of the radiating field, and therefore produce a specific intensification of the temperature gradient.

of reports indicate this figure. Therefore, the level of 10 mw/cm² can be considered safe only when it is continually checked at the body surface of a person.

References

1. Saidman-Meyer, J., *Les Ondes Courtes en Therapeutique*, Doin, Paris, 1956.

2. McLennan, J. C., "The Heating Effect of Short Radio Waves," *Arch. Phys. Therapy, 12,* 143 (1931).

3. Krasny-Ergen, W., "Mechanische Wirkungen der Kurzwellen," *Proc. 1st Intern. Congr. Shortwaves,* Vienna, 1937.

4. Groag, P., and Tomberg, V., "Zur Kurzwellentherapie," *Wien. klin. Wochschr., 46,* 929, 964 (1933).-

5. Groag, P., and Tomberg, V., "Zur biolog. Wirkung kurzer elektr. Wellen," *Wien. klin. Wochschr., 47,* 267 (1934).

6. Heller, R., "Lokalisierte Durchwärmung mittels Ultra-Kurzwellen," *Z. Ges. exptl. Med., 83,* 299 (1932).

7. Tomberg, V., "Die spezifischen biolog. Wirkungen kurzer elektr. Wellen," (a) *Proc. 1st Intern. Congr. Radio Biology,* Venice, 1934; (b) *Proc. Ges. Kurzwellenforschung,* Vienna, Nov. 1934.

8. Tomberg, V., "Bases Scientif. et Conceptions Nouvelles de l'Utilisation des Ondes Courtes," *Acta Physiother. Rheumat. Belg., 4,* 109 (1947).

9. Tomberg, V., "L'Effet Destructif de Micro-Ondes en Biologie," *Proc. 1st Intern. Congr. Medical Electronics, Acta Physiother. Rheumat. Belg.*, *6*, 295 (1948).

10. Tomberg, V., "Bio-negative Actions of Micro-Waves," *Proc. 12th Ann. Conf. Electrical Techniques in Medicine and Biology*, Philadelphia, Nov. 1959.

11. Tomberg, V., "Biological Microwave Hazards," *I.R.E. Convention*, New York, March 1960.

Microwave Radiation in Relation to Biological Systems and Neural Activity

J. Fleming, Jr., L. Pinneo,* R. Baus, Jr.,
and R. McAfee
Tulane University
New Orleans, Louisiana

INTRODUCTION

Neural and general physiological effects of microwave radiation are being studied through the Biophysics Program of Tulane University with the sponsorship of the Rome Air Development Center and the Office of Naval Research.

Commencing with electrical measurements and studies conducted several years ago, the investigative program at Tulane has more recently become centered about experiments at microwave power absorption levels below those resulting in evident pathology. The present study areas may be categorized as follows:

a. Investigation of the nociceptive response to microwave radiation using temperature, blood pressure, and electropotential measurements in conjunction with histopathological observations. This is a continuation of work reported last year and is to be presented by Dr. Robert McAfee of the Physiology Department of Tulane University, and the Veterans Administration Hospital in New Orleans.

b. Study of integrated neural activity using central nervous system d.c. potential measurements. The major portion of the d.c. work has been carried out by Lawrence R. Pinneo, now a research fellow with Giuseppe Moruzzi's group in Pisa, Italy.

---

* Formerly of McGill University, presently at University of Pisa.

c. Codification of electrical properties of tissue including absorption and reflection parameters for various microwave frequencies and consideration of the relationship of imposed electric and magnetic fields.

d. Development of a system for the remote determination of tissue surface temperature. This design is by Rene Baus of Tulane University.

INTEGRATED ACTIVITY AND DC POTENTIALS IN THE CENTRAL NERVOUS SYSTEM

Several of the groups investigating the biological effects of microwave radiation have recently made observations indicating changes in integrated neural activity due to microwaves. A most startling report on neural effects was presented at last year's conference by Bach (1) in which marked generalized states of arousal and drowsiness were observed, and similarities to the effects associated with direct electrical stimulation were noted. At the same conference, the University of California investigators reported what is possibly a conditioning effect on the thermal regulatory system of mice by radiation (3 cm, 156 mw/cm^2), and the University of Miami group described the cumulative depression of a learned response with hooded rats after being irradiated. Results of irradiation of dogs, with the beam directed toward the top of the head, were reported by the University of Iowa investigations. No evidence of central nervous motor excitation was observed. The University of Rochester reports that in their irradiation (10 cm, 165 mw/cm^2) experiments with dogs, there was a consistent increase in activity or agitation. In connection with the apparent conflict between the observations at Iowa and those at Rochester, Dr. Richardson of St. Louis University described the relative ineffectiveness of direct skull irradiation in heating the central nervous system.

To provide a quantitative measure of integrated activity for further microwave studies, the present investigators have developed techniques using the steady or direct current (dc) potentials of the central nervous system. These potentials, on the order of a few millivolts in magnitude, have been found to change with changes in integrated neural activity, such as arousal, and with changes in temperature. Our work in this area during the past year has been centered about:

a. Identification of causes and effects related to cortical dc potentials.

b. Development of appropriate surgical and electrical measurement techniques.

c. Correlation of dc potential changes and the nociceptive response observed with microwave irradiation.

Dc Potential Correlates

The work of previous investigators on cortical dc potentials has been described in a review by Pinneo (2). We shall presently describe both the use of applied potentials and passive measurements of dc potentials.

(*a*) *Applied Potentials.* The functional effects of predetermined dc levels may be studied by applying a potential from an external source using electrodes and polarizing the brain of a small organism during simple learning and perception. In this study, three experiments have been conducted, each experiment taking 2 days. On the first day whole brain polarization was carried out on 3 rats during bar pressing performance (in one) and simple maze performance (in the other two). On the second day local polarization of the visual cortex was employed on the same tasks. The same rats were used for three successive periods, and during each period the voltage of polarization was set at a higher value. The rats were of the hooded variety, taken at full maturity. Voltage of polarization in each case was adjusted by adjusting the electrode resistance to give a constant current of 10 μamp on the first period, 100 μamp on the second period, and 1000 μamp (1 ma) on the third period. Results on either task, for both local and whole brain polarization, and at the currents noted are presently being analyzed. Observations of the effects of the two lower currents are inconclusive, possibly because of maturity variations in the animals used. The 1 ma current, however, had a definite effect, causing arrest of task performance in most of the animals for the periods of stimulation though not apparently interfering with subsequent performance. The duration of the stimulation was varied from $\frac{1}{3}$, $\frac{2}{3}$, and the total time in the performance situation at three separate periods during a test day for each animal. The results indicate that the methods employed in this experiment might be useful for more extensive studies to characterize dc potential correlates and to provide a basis for interpretation of any polarization effects which might be found with microwave absorption.

(b) *Power Measurement.* These experiments were designed to determine the amount of electric power deliverable by the cortical dc generators as a function of various locations and to determine the laws of combining more than one set of electrodes taken in series and in parallel. Other experiments, associated with these studies, were designed to load the generator with discrete external loads during certain functioning processes (e.g., learning). In addition, the resistance effects of the skin, muscle, bone, dura, etc., overlying the cortex was to be ascertained relative to the power generators. None of these experiments have been completed, however, due to the problem of finding a suitable electrode system. This problem is discussed more fully later in this report.

(c) *Measurement of dc Potentials on the Human Scalp.* Attempts to measure dc potentials from the active human brain via scalp electrodes, though successful in eliminating artifacts due to changes in the potentials of the skin, proved to be more sensitive to the retinal-occipital potential of the eyes than to any underlying dc potentials from the cortex. The cortical potentials could be obtained, however, by using fixation techniques for maintaining the eye in one relative location and presenting moving stimuli to each subject such that it passed in front of the eyes. There were definite potential shifts of the order of magnitude reported by Kohlmer (about 400 μv). These changes could not be measured during changes in states of arousal because of the large shifts in retinal-occipital potential that also occurred with changing activation.

(d) *Regional or Cytoarchitectural Correlates.* A set of experiments was originated to correlate resting dc potentials and changing dc potentials with regions of the brain categorized in terms of function and cellular distribution patterns. Four conditions of measurement were contemplated for the whole brain and for prepared brain sections measuring both the resting potentials and the changing potentials: (1) without stimulation; (2) with stimulation of receptors; (3) with stimulation of relay nuclei and control centers, including the cochlear nucleus, the reticular formation, and the mid-line thalamic nuclei; and (4) with chemical and mechanical stimulation. Data have been analyzed for the resting potentials only.

Resting dc potentials are to be distinguished from the changing dc potentials. The former are relatively large potentials in the millivolts region that seldom change with usual types of neural stimulation. The resting potentials are subject to some changes, however, and since they also contribute to the field in the volume conductor that determines the activity of neurons, they are quite important. The "changing dc potentials" on the other hand, are the usually slight (200 to 500 μv) changes, occurring with specific direct or indirect neural stimulation. Ordinarily these potentials can only be measured when the resting potential just described is "bucked" out with a balancing potentiometer placed in series with one of the electrodes.

Brain regions with specific functions and cellular distributions were chosen for electropotential measurement. Figure 1-1 shows a schematic of the surface of the cat cortex and the functional areas as suggested by Campbell in 1905 and since corroborated. The location of the measurement electrode is indicated by number. For example, in the tables to follow, L1 refers to the left motor cortex, whereas R6 refers to the right auditory area.

Cats weighing approximately 2.5 kg were selected and anesthetized with 25 mg/kg of nembutal, given interperitoneally. After surgical preparation, electrophysiological measurements were made using calomel half cells with normal saline as the salt bridge. The

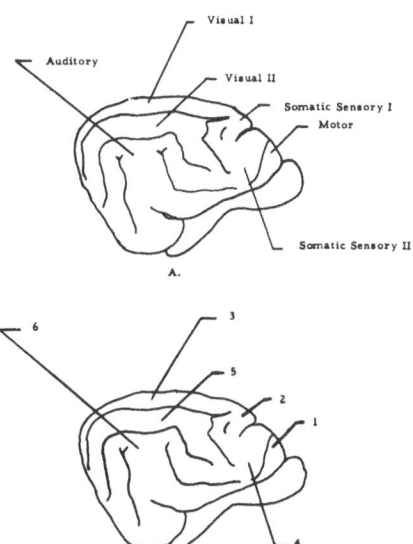

FIG. 1-1. Schematic of the surface of cat cerebral cortex showing the functional areas cytoarchitecturally (after Campbell, 1905) and the location of 6 measurements of dc resting potential.

contact electrodes were cotton wicks, saturated in the NaCl. The electrical reference was the periosteum in 4 of the 5 cats reported upon here, while the ventricle was the reference for the fifth. The electrodes were led to a Hewlett-Packard 425A Microvolt-ammeter, which in turn drove a 1000 ohm Esterline-Angus 1 ma recorder. Before and after each potential measurement from the cortex, the electrode open circuit voltage was measured to detect extraneous polarization. Calomel cells chosen had less than 1 mv potential difference among them at the outset.

A typical order of measurement was as follows: following the control measurement, the wick of the active electrode was placed against area R1. About 12 mm^2 of electrode was maintained in contact with cortical tissue at all times. Both potential and current measurements were taken for each location in numerical sequence followed each time by an electrode control measurement. After measurements were taken of all regions from R1 to L6, the sequence was repeated. During this entire procedure, the animal was intermittently given glucose and saline intravenously. Following all measurements, the animal was sacrificed by an injection of a lethal dose of nembutal; the brain was perfused and removed for histological examination.

Results on the resting potential measures as a function of area and time are shown in Table I. The periosteum and ventricle references are kept separate. Open circuit voltage and current flow between the two electrodes are shown. Electrode resistance was approximately 10 k-ohms.

The potential and current measurements indicate that the resistance between the electrodes, that is, the internal resistance of the biological generator, ranges between 3000 and 15,000 ohms. Such low resistance generators indicate that it is practically useless to compare the dc potential between different areas unless the resistance or current flow is also taken into account. Assuming the electrode resistance to be equal to 10 k-ohms for all measurements, this value was subtracted from the total resistance to obtain the generator resistance R_G. A new value for generator current, I_G, was then calculated using the measured potential, E.

Assuming a regional electrical similarity, the mean of the first and second measurements for all areas for E and I_G were calculated. Their product, EI_G, was then calculated as a useful parameter for comparison purposes. It may be noticed that the values for the power

TABLE I

Mean dc Voltage and Current Measured from Twelve Cortical Areas as a Function of Time

(ca. 45 Minutes)

Electrode Position	Ventricle Reference				Periosteum Reference			
	First Measurement		Second Measurement		First Measurement		Second Measurement	
	E	I	E	I	E	I	E	I
R_1	−0.2	.06	+1.00	.30	−4.9	.10	−1.6	.40
R_2	+1.0	.40	+0.10	.90	−4.1	.10	−5.4	.30
R_3	+1.4	.40	+0.20	.06	−4.5	.18	−4.6	.20
R_4	+1.2	.20	+1.00	.30	−3.7	.18	−4.2	.20
R_5	+0.5	.30	−0.80	.04	−5.3	.26	−5.6	.30
R_6	+0.4	.08	−0.10	.02	−4.9	.26	−5.4	.30
L_1	+1.1	.30	+1.50	.50	−2.3	.10	3.8	.28
L_2	+0.5	.15	+0.20	.10	−4.3	.20	−6.4	.38
L_3	+0.8	.15	+0.30	.14	−5.9	.32	−7.4	.44
L_4	+0.2	.12	−0.50	.25	−3.7	.20	−4.8	.22
L_5	+0.1	.40	+0.50	.03	−6.3	.34	−7.4	.42
L_6	+0.2	.40	+0.30	.08	−6.3	.38	−7.4	.44

E is in millivolts, I in microamperes.
Sign refers to active electrode (for both E and I, though only shown in E).
E and I represent values after voltage and current due to electrodes have been subtracted.
Mean electrode potential before measurement was 800 microvolts; current 0.005 microamperes.
Mean electrode potential after measurement was 1.2 millivolts; current 0.15 microamperes.

capacity of the biological generator are similar to those obtained last year. Results of these calculations are given in Table II (only the periosteum reference was calculated).

A pattern of power variation may be observed in both right and left hemispheres, in which the auditory areas tend to have higher values than those found in other areas. A more adequate interpretation may be made by including the data on the changing potentials. A more conclusive observation can be made regarding the resting potentials. These potentials are not a membrane phenomenon, as might be supposed because of the layer of pia matter overlying the cortex, since the low internal resistance for all regions measured is several log units lower in magnitude than that found in the membrane.

TABLE II

Mean dc Parameters Measured from Twelve Cortical Areas with Reference to Periosteum and Corrected for Electrode Resistance

Electrode Position	E_{cc}(mv)	I_G(μamp)	R_G(k-ohms)	P_G(10^{-9} watts)
R_1	− 3.25	1.090	3.00	3.53
R_2	− 4.75	0.346	13.75	1.64
R_3	− 4.55	0.325	14.00	1.48
R_4	− 3.95	0.366	10.08	1.44
R_5	− 5.45	0.576	9.45	3.14
R_6	− 5.15	0.614	8.40	3.16
L_1	− 3.05	0.500	6.10	1.53
L_2	− 5.35	0.630	8.50	3.37
L_3	− 6.65	0.888	7.50	5.90
L_4	− 4.75	0.389	12.20	1.85
L_5	− 6.85	0.867	7.90	5.94
L_6	− 6.85	1.160	5.90	7.86

Sign refers to active electrode
Electrode resistance assumed to be 10 k-ohms in all cases

Additional data on the function of the resting potentials has been obtained by observing changes due to anoxia and due to death with nembutal poisoning. With anoxia, induced by tracheal clamping, the potentials (measured with respect to the periosteum), diminish with time, returning to their original value upon release of the clamp. In contrast, in response to a lethal dose of nembutal, the potentials increase, with the increase persisting for several hours, never (apparently) returning to zero. A similar (positive going) increase occurs with the tracheal clamping experiments within a short time after death.

With completion of the analysis of both resting and changing steady potential data for various stimuli, the extent of neuronal participation in each can be estimated.

Development of Techniques

(a) *Development of Special Surgical Techniques.* Cortical steady potential measurements require surgical techniques which insure a nonshock, nontraumatized brain preparation for long-term experiments. To protect the brain from injury due to drying out or becoming cold, a method was developed (after Marshall and Perot) to employ the loose and opened scalp as a protective fluid well over the exposed cortex. This reservoir is filled with either Ringer's solution or mineral oil at body temperature, and the solution is changed periodically during an experiment. In order to keep the animal out of shock, femoral cannulation and administration of glucose are employed periodically. Methods for preventing hyperventilation of cerveau isole preparations under artificial respiration were tested and perfected.

(b) *Electrode Tests and Construction.* The measurement of dc potentials in biological tissue requires a relatively stable electrode system. Potentials in biological tissue range from less than 1 to about 100 mv with currents from 1 to 6 μamp. Electrochemical interactions between most conductors and biological fluids produce electrical parameters of the same order.

An established technique of measurement is to employ a metal electrode in series with one of its salts and a salt bridge such as normal saline, Ringer's solution, or saturated KCl. The calomel half-cell used in pH measurement represents a refined version of this approach. These electrodes, though useful, are large in terms of physical size and resistivity (15–20 k-ohms). The large size is a compelling disadvantage in chronic or long-term measurements.

To explore alternatives, a smooth platinum electrode system was tested. Electrode voltages of the order of 300 mv with variations of 10–20 mv were observed. With a bucking circuit added to provide a balancing potential, excessive current with hydrogen bubble formation was observed. Attempts to buffer were not successful.

The use of conventional silver-silver chloride electrodes involved the problem of obtaining a sufficiently low resistance with excessive electrode surface. The insertion of large surfaces gives rise to large masking currents of injury.

An attempt to employ an ac system using a relay interruptor and capacitative coupling from the electrodes was not successful due to its complexity and susceptability to ambient interferences.

A recently developed silver-silver chloride wafer electrode with promising characteristics has been tested. Measurements in saturated sodium chloride indicated resistance of 2–4 ohms. Measurements in 0.1 normal, normal, and saturated sodium chloride show independence from the concentration of the electrolyte.

Over a period of 48 hr, the self-polarization potentials of these electrodes changed by less than 50% of the original values, which were approximately 800 μv. Polarization obtained by applied currents of 10, 100, and 1000 μamp returned to a minimum value in less than 10 min. In general these electrodes appear to be adequate for further work on chronic cortical dc potential measurements.

Correlation of Cortical dc Potential Changes and the Nociceptive Response

The nociceptive response is to be described subsequently by Dr. Robert McAfee. It is an avoidance or withdrawal response associated with heating of peripheral nerves to about 45°C. The relation between the response and dc potentials has been investigated experimentally.

In these experiments, the cat is surgically prepared as described in the previous section. The sciatic nerve is exposed and irradiated with concurrent measurements of dc potentials in the left sensory motor cortex.

In the first experiment, the nociceptive response was obtained on three occasions. Voltage measurements were made during the first and last response periods and current was measured during the second period. Temperature was measured at the sciatic nerve.

Results of the first experiment are shown in Figure 1-2. The changes in cortical potentials are of the order of 1 mv, and con-

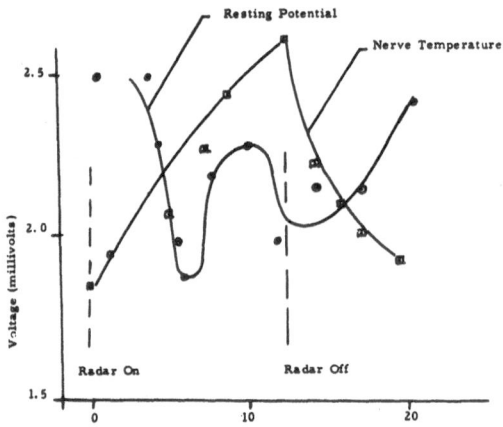

Fig. 1-2. Temperature and dc cortical potentials with 3 cm radiation.

sistently decrease with increasing temperature. This is confirmatory to earlier experiments with both microwave and infrared radiation. In all experiments, the dc changes are relatively small, indicating that the extreme changes in blood pressure and limb position found in the nociceptive response do not cause equivalent effects on the steady potentials of the cortex. This observation must be tempered to await additional experimental data and assessment of the effect of choice and depth of anesthesia.

ELECTRICAL PROPERTIES OF TISSUE

Reflection and Absorption of Microwave by Tissue

The present section extends the scope of reflection and absorption data presented at the Microwave Conference in August 1959 (1). Data for depth of penetration, wavelength in tissue, and reflection coefficient for certain tissue interfaces are given for a range of frequencies from 100 to 35,000 mc for various tissues in Tables III–V. Dielectric data employed in the related calculations are from the work of Schwan (3), England (4), and Baus (5).

In our previous report, an equation was provided for power density computation:

$$I_x = I_0(I - R_{1,2}) \exp(-x/D_2)$$

in which I_x = the power density in the second medium (more distant from the source) at a distance x from its interface with the first medium (nearer the source),

TABLE III

Depth of Penetration for Various Frequencies (in cm)

	100 mc	200 mc	400 mc	1000 mc	3000 mc	10,000 mc	24,500 mc	35,000 mc
Yellow Marrow	22.9	20.66	18.73	11.9	9.924	0.34	0.145	.0730
Brain	3.56	4.132	2.072	1.933	0.476	0.168	0.075	.0378
Lens	9.42	4.39	4.23	2.915	0.500	0.174	0.0706	.0378
Vitreous Humor	2.17	1.69	1.41	1.23	0.535	0.195	0.045	.03145
Fat	20.45	12.53	8.52	6.42	2.45	1.1	0.342	—
Muscle	3.454	2.32	1.84	1.456	—	0.134	—	—
Whole Blood	2.86	2.15	1.787	1.40	0.78	0.148	0.0598	.0272
Skin	3.765	2.78	2.18	1.638	0.646	0.189	0.0772	—

TABLE IV

Wavelength in Tissue for Various Frequencies (in cm)

	100 mc	200 mc	400 mc	1000 mc	3000 mc	10,000 mc	24,500 mc	35,000 mc
Yellow Marrow	116.1	62.2	32.19	12.63	3.97	1.250	0.368	.388
Brain	31.7	19.4	11.16	4.97	1.74	0.595	0.200	.201
Lens	33.15	22.3	12.53	5.28	1.75	0.575	0.200	.201
Vitreous Humor	21.7	13.0	7.96	3.41	1.18	0.395	0.146	.154
Fat	96.0	57.1	30.9	12.42	3.79	1.450	0.68	—
Muscle	27.65	16.3	9.41	4.09	—	0.616	—	—
Whole Blood	25.15	15.35	8.89	3.87	1.36	0.449	0.214	.1675
Skin	28.07	17.94	10.12	4.41	1.49	0.506	0.25	—

I_0 = the power density in the first medium incident on the common interface,

$R_{1,2}$ = the reflection coefficient of the interface, and

D_2 = the depth of penetration in the second medium.

This equation may be useful for approximate calculations but it neglects standing waves, which could occur in particular situations. For example, if there is low absorption (i.e., deep penetration) for a particular wavelength-tissue combination, and if the thickness of the tissue is about one-fourth (¼) the wavelength, essentially no reflection will occur and there will be maximum transmission of energy into the layer below the ¼ wavelength tissue. Using the skin-air interface with normally incident 3 cm radiation, as a useful case, we find that the wavelength in skin is 0.51 cm and thus a ¼ wavelength layer of skin would have a thickness of 0.13 cm. For somewhat thicker or thinner skin layers, about 50% of the incident radiation is reflected at the skin-air interface but for the ¼ wavelength thickness the re-

TABLE V

Reflection Coefficients for a Sequence of Tissue Interfaces at
Various Frequencies (Air-Skin-Fat-Muscle)

	100 mc	200 mc	400 mc	1000 mc	3000 mc	10,000 mc	24,500 mc	35,000 mc
Air-Skin	0.758	0.684	0.623	0.57	0.55	0.53	0.47	—
Skin-Fat	.340	.277	—	.231	0.19	0.23	0.22	—
Fat-Muscle	.355	.3515	.3004	.2608	—	—	—	—

flection is reduced, though not eliminated since much of the power is absorbed by the skin. The power not absorbed may be calculated using the equation above for I_x. Based on this calculation, about ⅓ of the power entering the skin would appear at the distant interface to increase transmission.

The listing of all the calculated reflection coefficients for the various tissues is too extensive a procedure for the present conference. A tabulation of reflection coefficients for the air to skin, the skin to fat, and fat to muscle interfaces is given in Table V.

The values of certain of these parameters are useful in interpreting the results reported in various animal experiments. In particular, it is notable that the skin-air interface is a remarkable reflector at the longer, more penetrating, wavelengths. At 100 mc, the skin reflects 76% of the incident power and, though this value decreases, 57% of power is reflected at 1000 mc, and 47% at 24.5 kmc.

Experiments with Fields

The use of large electric and magnetic fields in experiments with animals or simple tissues offers other approaches to the study of non-thermal electromagnetic effects than are available with conventional microwave irradiation techniques.

The pilot experiments conducted in this area includes:

a. Measurement of dielectric saturation using large electropotential fields.

b. Neural stimulation in rapidly changing magnetic fields.

c. Observations of reradiated power.

The dielectric saturation measurements depend on the construction of a suitable cell. All attempts in this direction using a concentric cylindrical cell have been unsuccessful due to a breakdown across the cell elements. This experimental approach appears promising, however, and a new cell with flat parallel plates is being designed.

Magnetic stimulation experiments using the sciatic nerve muscle preparation indicated no stimulation or change in threshold.

The possibility of reradiation at the pulse repetition rate has been mentioned previously. The study of reradiation possibilities requires the development of an extremely sensitive detector. Attempts with a pick-up coil around the body of a mouse and with external shielding gave no indication of reradiation in the course of experimentation.

A RADIOMETER FOR REMOTE MEASUREMENTS
OF SKIN TEMPERATURE*

The problem of accurately measuring the surface temperature of
an animal located in a microwave irradiation field, long a topic of
interest among investigators studying the biological effects of microwaves, is perhaps best solved by use of a radiometer that is capable
of remote surface temperature measurements from a vantage point
located outside of the interfering effect of the field.   That such a device is passive and does not require physical contact with the animal
under observation implies that temperature measurements can be
made without seriously perturbing either the microwave irradiation
field or the temperature distribution of the animal.   This is patently
not true of temperature measurement devices such as thermometers,
thermocouples, and thermistors, which do require intimate physical
contact with the animal and which do significantly interfere with
both the microwave field and the temperature distribution of the
animal.

A radiometer is an optical device which collects a part of the
radiation emitted by the surface of a body and focuses it onto a detector which in turn provides an electrical signal output that is a
known function of the collected radiation.   A meter or a recorder
with a suitable scale can be designed to convert this electrical signal
into visual indications that relate to the temperature of the body.   A
design for an infrared radiometer is presented, indicating feasibility.

Description of the Radiometer

A schematic drawing of the radiometer is shown in Figure 2-1.
The major components are a cassegrainian reflector, a rotating shutter, a silicon filter, and an indium antimonide detector.   The radiation emitted by a portion of the surface to be measured is collected
by the mirror system and is focused onto the sensitive element of the
detector after passing through the rotating shutter and the optical
silicon filter.   The electrical output of the detector is then fed into a
tuned circuit and subsequently amplified to operate a recorder or a
meter.

* Designed by R. Baus.

The secondary diameter mirror shown in Figure 2-1 should ideally have a hyperbolic reflecting surface in order to reduce the spherical abberation to a minimum, but the main parabolic mirror obtained is not of sufficiently high surface quality to warrant such a refinement. In fact, image quality considerations for this system are much less exacting than photometric properties. To provide a useful arbitrary specification on the image quality of the optical system, the circle of least confusion in the image plane corresponding to a distant point source is to be less than 0.020 in.

Should it become desirable to improve the image quality, as might be necessary for a scanning device, we would design a cadadioptic meniscus system (with optical silicon in place of glass) along the lines suggested by Bouwens (6). The high refractive index of optical silicon (7) would permit the design of a combination refractive-reflective system having both high numerical aperture and excellent image quality.

The indium antimonide detector is located within an evacuated metal chamber or heat sink maintained at a constant temperature of 20°C by means of a Peltier microrefrigerator. This is a solid-state device for bringing about a temperature differential using electrical power. A slit aperture connected to the heat sink is provided to sharply limit the field of view of the optical system.

The rotating shutter, a disc with radial slots on its periphery, is

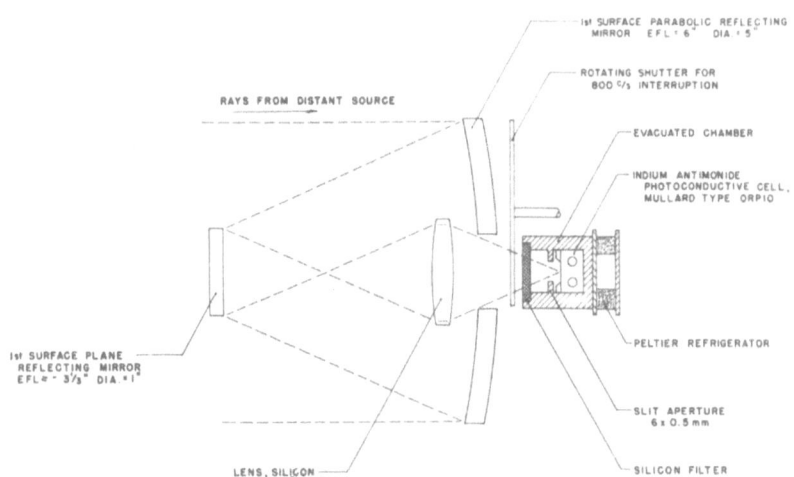

Fig. 2-1. Schematic drawing of infrared radiometer.

to be polished and silvered on the face adjacent to the detector. Thus, the detector sees alternately (at a rate of 800 viewing periods per second) the distant object at a temperature of T and a reflected image of itself at a temperature of 20°C; it measures the temperature difference $\Delta W = T - 20°C$.

The Signal to Noise Ratio

The signal to noise ratio is the basic measure of the detectivity of an infrared radiometer; that is, the precision of the measurement is noise limited.

The relationship between the precision of the measurement and the signal to noise ratio is, however, dependent on the time over which the measurement takes place. For example, the signal to noise ratio of a coherent signal in an ambient of incoherent noise can be markedly improved by integrating the signal over a long time base. But electronic integrators of the type required for this purpose are costly and elaborate, and presumably out of place in a device of this type which ideally should present a visual indication of temperature with a delay of at most a few seconds. For this reason, the minimum usable signal to noise ratio (SNR) is arbitrarily set at 3:1, a value that can be easily handled by conventional electronic circuits and recorders.

In order to calculate the accuracy of the radiometer, assuming that the system beyond the detector can measure with an SNR of 3:1, we need to know both *the change in the signal power corresponding to a change in the temperature of the source and the noise equivalent power of the radiometer*. The value of the signal power can be found through use of the laws of black body radiation and the data of the optical system; the value of the noise power is provided by data for the noise contributions of the components of the radiometer, particularly noise for the detector.

The Noise Equivalent Power of the Indium Antimonide Detector

The signal to noise ratio of the indium antimonide ORP 10 detector (International Electronics Corp.) has been measured using a known amount of interrupted monochromatic radiation (8). With 2 μw of 6.0 μ radiation interrupted at a frequency of 800 interruptions or cps, the signal to noise ratio is in the neighborhood of 80 with the detector at 293°K.

The spectral response of the detector is shown for the wavelength band 1 to 7 μ in Figure 2-2 with a black body curve for 305°K (32°C). The change in output voltage of the detector increases almost linearly with wavelength within the range of from 1 to 7 μ and at 7 μ it drops off rather sharply to zero. If we define a figure of merit M to be the ratio of the response at a wavelength of λ to the response at 6.5 μ, for the present detector:

$$M = -0.05 + 0.156\lambda,$$

with λ given in microns.

Using the signal to noise ratio obtained above for a particular wavelength and a particular power, together with the expression for M, one is able to calculate the signal to noise ratio (SNR) for signal of any wavelength and power.

In order to calculate the SNR, it simply remains to calculate the signal from the source.

Black Body Radiation

From Planck's radiation equation (9), giving the radiation density per unit wavelength interval in an isolated thermal enclosure, one may obtain an expression for the spectral emittance, B_λ of a radiating surface with unity emissivity.

$$B_\lambda = \frac{1.19 \times 10^8}{\lambda^5}\left[\exp\left(\frac{1.44 \times 10^4}{\lambda T}\right) - 1\right]^{-1}\frac{\text{watts}}{\text{steradian-meter}^2\text{-micron}}$$

where λ is expressed in microns and T is expressed in °K.

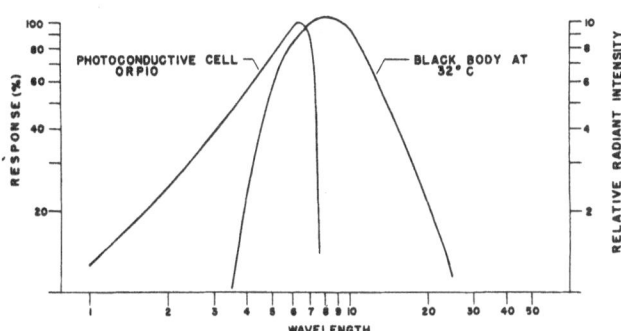

FIG. 2-2. Spectral relationships for the measurement of body temperature with indium antimonide detector.

Since, for the range of temperatures and wavelengths of interest in this discussion $(1.44 \times 10^4)/T \simeq 950$, the following approximation can be made

$$B_\lambda = \frac{1.19 \times 10^8}{\lambda^5} \exp\left(\frac{-1.44 \times 10^4}{\lambda T}\right) \frac{\text{watts}}{\text{steradian-meter}^2\text{-micron}}$$

The Effective Emittance of the Source

Since the radiometer is not equally sensitive to different wavelengths and since the wavelength bandpass of the radiometer is limited, it is necessary to define an effective emittance of the source. By effective emittance, we mean the emittance of an ideal 6 μ monochromatic source that would produce the same radiometer output as would the actual source.

The spectral response of the radiometer is primarily determined by two of its components: the silicon filter and the idium antimonide detector. The silicon filter has a sharp cut off at 1 μ and transmits with essentially a flat response in the range of from 1 to 7 μ. The spectral response of the detector has been discussed above. Thus, the spectral band pass is limited to the 1 to 7 μ range.

Furthermore, since the reflectivity of the aluminum coatings on the reflecting surfaces of the mirrors changes only slightly in the bank 1-7 μ, it follows that the spectral response of the radiometer is given by

$$M = -0.05 + 0.156\lambda.$$

Therefore, the effective brightness of the source is given by

$$B = \int_{\lambda=1}^{\lambda=7} B_\lambda M d\lambda.$$

If this expression is integrated by parts,

$$B = -(5.95 \times 10^6)\left(\frac{T}{\lambda^3 N} + \frac{3T^2}{\lambda^2 N^2} + \frac{6T^3}{\lambda N^3} + \frac{6T^4}{N^4}\right)\exp(-N/\lambda T)\Big|_{\lambda=1}^{\lambda=7}$$

$$+ (1.855 \times 10^7)\left(\frac{T}{\lambda^2 N} + \frac{2T^2}{\lambda N^2} + \frac{2T^3}{N^3}\right)\exp(-N/\lambda T)\Big|_{\lambda=1}^{\lambda=7}$$

where $N = 1.44 \times 10^4$.

If B is differentiated with respect to T, and if T is assumed to be in the neighborhood of 300°K, then it is found that

$$dB/dt = 0.24 \text{ w/°K-steradian-meter}^2$$

which gives the change in effective emittance of the source for a change in the source temperature of $1°K$.

The Photometry of the Optical System

For the purpose of this calculation, the optical system shown in Figure 2-1 may be represented by a lens having an effective focal length, f_1 and a clear aperture of A. Power emitted by an incremental area of the source is gathered by the lens and focused onto the area lying in the plane of the sensitive surface of the indium antimonide detector.

If the source is assumed to be an isotropic radiator, the change in the flux with temperature at the sensitive surface of the detector is given by

$$dF/dT = (\pi K_1 dS/4N^2)\,(dB/dT)$$

where dB/dt = change in effective emittance per unit $°K$ of the source,

K_1 = transmission coefficient of lens,
A = diameter of lens, and
N = relative aperture of lens = f/A.

For the optical system shown in Figure 2-1 the following values may be assumed:

$K_1 = 0.81$,
$A = 5$ in.,
$f = 6$ in., and
$dS = 3 \times 10^{-6}$ m² (i.e., the sensitive area of the indium antimonide detector).

Using these values together with the value of dB/dT obtained previously, one obtains

$$\Delta F/\Delta T = 0.32 \ \mu w/°K.$$

In other words, the flux incident on the detector changes by 0.32 μw for a $1°K$ change in the temperature of the source.

Conclusion

The signal to noise ratio associated with a 2 μw signal was shown to be about 80:1 for the particular detector chosen. For a 0.32 μw signal (a $1°K$ change) the signal to noise ratio would be 12:1.

Using the criterion that a SNR of 3:1 can be detected easily, it follows that the radiometer should be able to measure a change of ¼°K.

The infrared sensitive element is rectangular in shape (6 × 0.5 mm) and would measure an area 12 cm × 1 cm at a distance of 10 ft. If temperature changes of less than 1°K can be ignored, the sensitive element would be masked to measure an area 3 cm × 1 cm. Alternately, the detection system could be improved to measure at smaller signal to noise ratios so as to measure smaller temperature changes.

A number of methods can be used for improving the precision of measurement:

1. A lower heat sink temperature. This will reduce the noise component resulting from "Generation-Recombination Noise by Lattice Excitation and Recombination" (10). It will also reduce the background radiation noise. Moreover, it should increase the SNR by virtue of increasing responsivity (11) of the detector.

2. Electronic integration of the signal over a long time base. This may permit using a SNR as low as 0.01.

3. Use of a narrow band pass amplifier for the purpose of filtering the excess noise.

4. The use of an optical system having a larger relative aperture.

References

1. Bach, S. A., "Some Effects of Ultra-High Frequency Energy on Primate Cerebral Activity," *Proc. Third Ann. Tri-Service Conf. on Biological Effects of Microwave Radiating Equipments,* 82, August, 1959.

2. Pinneo, L. R., "Direct-Current Potentials of the Central Nervous System," Technical Report RADC-RN-59-137 ASTIA Document No. AD-214692 Rome Air Development Center, Air Research and Development Command, Griffiss Air Force Base, N.Y., June, 1959.

3. Schwan, H. P., "Survey of Microwave Absorption Characteristics of Body Tissue," *Proc. Second Ann. Tri-Service Conf. on Biological Effects of Microwave Energy,* 8, 9, 10, July 1958, p. 126. Astia Document No. AD 131477.

4. England, J., "Dielectric Properties of the Human Body for Wavelengths in the 1–10 cm Range," *Nature, 166,* 4220 (September 16, 1950).

5. Baus, R., "A Method of Measuring the Complex Dielectric Constant of High Loss Materials in the Range of Centimeter Waves," Annual Progress Report (1955–1956) December 17, 1956. Contract Nonr-475(03), Office of Naval Research.

6. Bouwens, A., *Achievements in Optics,* Elsevier, Amsterdam, 1946.

7. Dow Corning Corporation, Technical Bulletin 2-305, May, 1960.

8. ORP 10, Photoconductive Cell, Instruction Manual, issued by Mullard, Ltd., Undated, New York, New York.

9. Planck, M., *The Theory of Heat Radiation,* Dover, New York, New York, 1959.

10. Petritz, R. L., "Fundamental of Surface-Red Detectors," *Proc. I.R.E.,* 47, 1458, September, 1959.

11. Bell, E. E., "Radiometric Quantities Symbol of Units," *Proc. I.R.E.,* 47, 1432, September, 1959.

Neurological Effect of 3 cm Microwave Irradiation*

ROBERT D. MCAFEE, CAROLYN BERGER
AND PHILIP PIZZOLATO
Radioisotope Laboratory and Clinical Laboratory of the
Veterans Administration Hospital, New Orleans
and the Biophysics Laboratory of Tulane University
New Orleans, Louisiana

INTRODUCTION

IN A PREVIOUS REPORT (1) we have described the neurophysiological response of decerebrated and anesthetized cats to locally applied 3 cm microwave radiation. Since this report, further work has confirmed and extended our original observations.

The response to locally applied 3 cm microwave radiation is nociceptive in nature and includes withdrawal movements, elevation of blood pressure, and respiration changes. The microwave energy which produces this response may be applied to a sensory nerve bundle such as the sciatic or radial nerve exposed for irradiation or to an area of skin rich in cutaneous nerve fibers. The response occurs when the temperature at the nerve or within the subcutaneous tissue reaches about 45°C. A nociceptive response in an unanesthetized but decerebrate animal includes reflex withdrawal movements from the noxious stimuli, a sharp rise in blood pressure, and alterations in rate and depth of respiration. The blood pressure response occurs in (nembutal) anesthetized animals to a lesser extent and the reflex withdrawal is generally inhibited.

Two important questions may be asked with regard to this phe-

* This work was supported by the U.S. Air Force Air Research and Development Command, Rome Air Development Center, Rome, N.Y.

251

nomenon: (1) Is the neurophysiological response observed a result of raising the temperature of sensory nerves to 45°C or a result of some nonthermal microwave effect? (2) Are the nervous and supporting tissues involved at the irradiated site damaged?

Much of the evidence which indicated that the neurophysiological effects of microwave radiation are nonthermal in nature resulted from animal experiments in which the head of an animal was irradiated and the subsequent behavioral changes which followed were compared with behavioral changes which then failed to occur when the back of the animal was irradiated (2). As a result of observations of this kind and others (3–5), it was presumed by some that the neurophysiological effect of microwave radiation was of central nervous system origin as a result of some nonthermal mechanism.

Evidence which we will present convincingly indicates that it is thermal stimulation of the peripheral nervous system which produces the neurophysiological and behavioral changes observed.

Results

It has previously been observed by C. von Euler (6) that warming exposed peripheral sensory nerves, such as the sciatic or radial nerve, to 45°C produces a nociceptive response in a decerebrate animal. We have repeated these experiments in decerebrate and anesthetized animals. The animals were decerebrated so that they would remain quiet but still be capable of exhibiting spinal reflexes such as the crossed extension reflex.

Figure 1 illustrates the thermode used to warm the exposed sciatic or radial nerve of a decerebrated cat. The nerve is placed within the groove of the thermode and warm water is circulated within the instrument. Blood pressure is measured by carotid or femoral artery cannulation; temperature, by a thermistor placed under the nerve. When the temperature at the nerve reaches about 45°C a nociceptive response is produced. The response is illustrated in Figure 2.

If the nerve is warmed by locally applied infrared radiation to about 45°C, a response identical to that produced by the thermode occurs. If a turn of electrically insulated resistance wire is placed around the nerve, the same response occurs at about 45°C; and if the nerve is irradiated with 3 cm microwaves, the identical response occurs when the temperature at the nerve reaches about 45°C. In each case the response is correlated with the attainment at the nerve of a temperature of 45 ± 2°C.

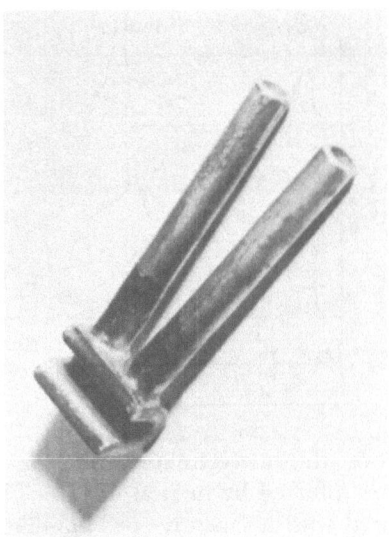

FIG. 1.   A thermode.

Figure 3 illustrates that in the case of the 3 cm microwave response, the effect is due to microwave induced heating *only,* since the response does not occur when the nerve is cooled with air while it is being irradiated.

Figure 4 illustrates that the effect is reversible and can be turned on and off as the nerve is repeatedly heated and cooled.   Figure 5 shows that the effect can be prolonged for at least 1 hr.

It has been shown by C. von Euler with the thermode (6) that it

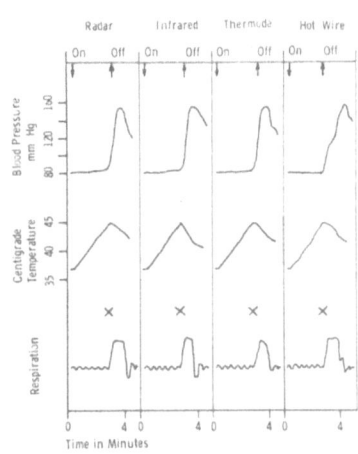

FIG. 2.   Stimulation of the sciatic nerve of a decerebrate cat.   The comparative effects of 3 cm microwaves (radar), infrared radiation, thermode, and hot wire are indicated.   The animal exhibited a crossed-extensor reflex at the time indicated by the symbol "✕."

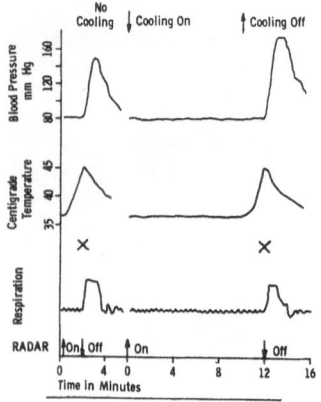

Fig. 3. Dependence of reflex production on temperature rather than on length of time irradiated. The animal exhibited a crossed-extensor reflex at the time indicated by the symbol "×."

is the thinly myelinated and unmyelinated sensory nerve fibers which are affected by heat at 45°C. These fibers are known to be associated with nociceptive or "painlike" reflexes. All peripheral cutaneous afferent nerve branches are supplied with these fibers. It seemed reasonable to us that 3 cm microwave radiation could penetrate to and heat these cutaneous nerves. Figure 6 shows their approximate location in an idealized section of skin and also illustrates the fact that for a part of their course these fibers lie within poorly vascularized subcutaneous fatty tissue. If the nerves were to be heated within this region they would tend to hold this heat and lose it to the capillary blood less rapidly than would nerves within the more richly vascularized dermis.

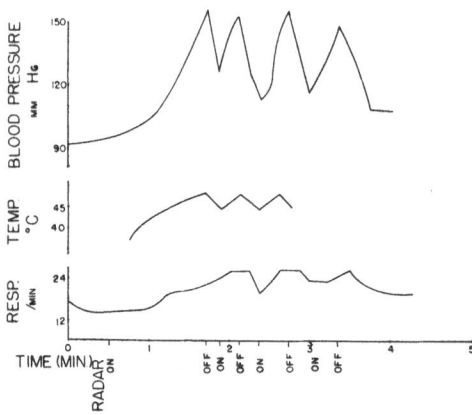

Fig. 4. Decerebrate cat, sciatic.
(July 24, 1959).

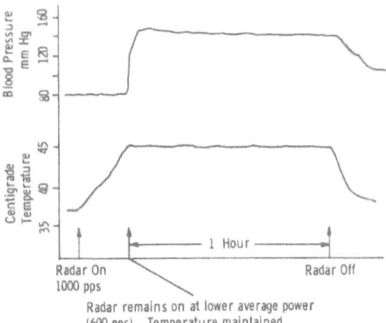

FIG. 5. Elevated blood pressure maintained with continuous micro-wave-induced heating. The cat was anesthetized with nembutal.

Figure 7 illustrates sections of skin peeled away from the animal and that the nociceptive response which occurred when these attached peelings were irradiated was in every way identical to those previously obtained. In addition, if skin from the paws, lower legs, face, and head of the animal were irradiated, without previous dissection of any kind, a strong nociceptive response occurred when the tempera-ture within the tela subcutania reached about 45°C. If the skin from the belly or back of the animal was irradiated, no response occurred even at higher temperatures up to 55°C. This result is not surpris-ing since the paws and head of an animal are richly supplied with cutaneous nerve fibers whereas the belly and back skin are very poorly supplied with cutaneous sensory nerve fibers (Figs. 8, 9).

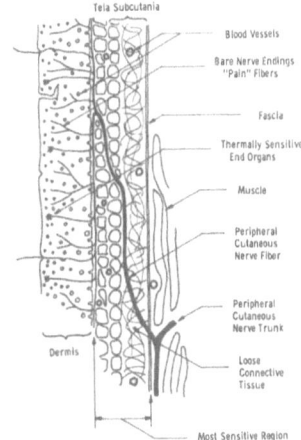

FIG. 6. Portion of peripheral cutaneous sensory nerve fiber lying in region where 3 cm microwaves produce the greatest temperature rise.

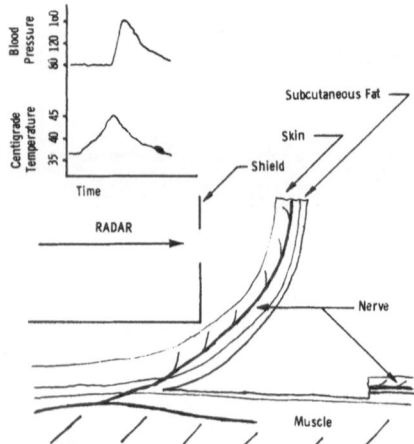

FIG. 7. Irradiation of skin flap.

These findings yield convincing evidence that the presumed non-thermal effect on the central nervous system is a result of thermal stimulation of peripheral sensory nerves. Now let us inquire if damage occurs to the nervous and supporting tissue involved at the irradiated site.

We have shown that the nociceptive response can be maintained for periods up to 1 hr if the temperature at the nerve or within the

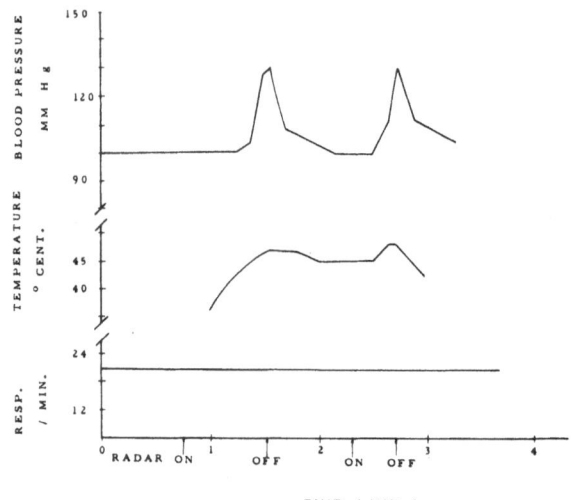

TIME (MIN)

FIG. 8. Decerebrate cat, forehead.

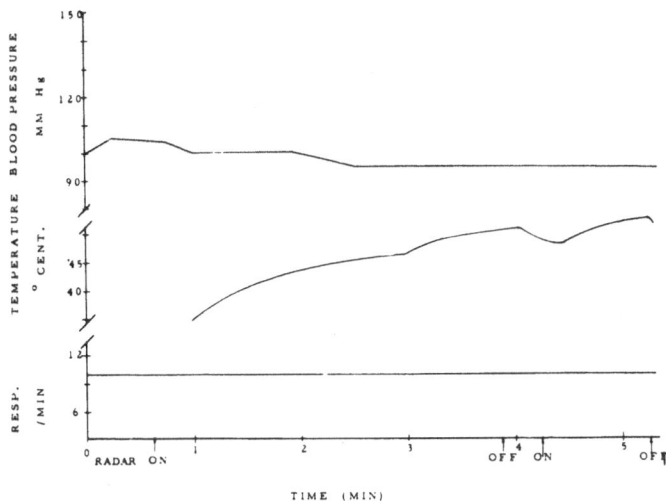

Fig. 9.   Decerebrate cat, abdominal skin flap.

tela subcutania is maintained at about 45°C.   This duration should be more than adequate to initiate histological changes should significant tissue damage occur.

Lower foreleg and lower hind leg skin of anesthetized cats were irradiated at 46–47°C for 1 hr periods.   Since histopathological damage may not become apparent for some time after a noxious agent has been applied, the experimental animals were sacrificed at intervals ranging from 3 hr to 2 weeks following exposure to microwave irradiation.   This time period allows for the appearance of fast and slow components of histological damage.

The histological findings are presented in Table I.   Stated briefly, they are as follows: The cutaneous nerves are unaffected by the irradiation although they are responsible for the nociceptive response. The vascular tissue is slightly but reversibly adversely affected by the radiation.   This damage may result wholly or in part from placement of a thermistor probe within the irradiated region for monitoring the tela subcutania temperature.   Tissue anoxia resulting from the prolonged local temperature elevation may also play a role in the vascular changes observed.   Needless to say, if the temperature is not controlled accurately and reaches 50°C or above, severe deep "burns" will result.

TABLE I

Preliminary Histopathologic Observations

Time after exposure (days)	Epidermis	Dermis	Vascular region	Nerve		
				Cell nuclei hematoxylin	Cyto-plasm eosin	Myelin sudan black
¼	Normal	Mild change in cyto-plasm (less eosinophilic —some basophilia degeneration)	Mild conges-tion related to general reddened appearance of skin	Normal	Normal	Normal
3	Thickened with some hydrophic degenera-tion or necrobiosis	Some cellular degeneration with basophilia	Congestion but no cellular infiltration or hemorrhage with continued reddened appearance	Normal	Normal	Normal
7	Normal	Mild basophilia	Congestion suggesting telangiectasis	Normal	Normal	Normal

Discussion

It is significant that the neural tissue appears to be unaffected by local temperature elevation to 46–47°C, and this suggests that the neural activity which results in the nociceptive response does not occur as a result of nerve injury. The nociceptive response is therefore a phenomenon which would ordinarily not be observed as a result of thermal radiation because the cutaneous blood flow would prevent the penetration of heat to these subcutaneous fibers and maintain their temperature at body temperature levels. However, 3 cm microwave radiation, because of its penetrating characteristics, is able to reach these cutaneous sensory fibers and heat them to about 45°C. Thus the nociceptive response is initiated and associated behavioral changes result.

Attempts have been made to characterize the nociceptive response in terms of other neural and humoral structures and mechanisms affected. These include sympathetic nervous system involvement and hormone release, ascending pathways (within the spinal cord) taken by the sensory impulse, brain stem reticular formation involvement, effects at the thalamic level, and central nervous system cortical involvement. Actually a great deal can be said about possible involvement of these structures if only by inference. We are confident that the radar effect is identical to the heat effect. From C. von Euler's work on thermal sensitivity of peripheral sensory fibers, we read that the specific fibers involved in the nociceptive response are the fibers of the C class and delta group. The afferent pathway taken by these fibers up to the thalamic level and into the cortex is well known, and the evoked response resulting from electrical thermal and mechanical stimulation of afferent nerves has often been described. We infer that the same pathways are taken and the same evoked responses elicited by microwave induced heating of afferent nerves as well.

References

1. McAfee, R. D., "Neurophysiological Effects of Microwave Irradiation," *Proc. of the Third Annual Tri-Service Conf. on Biol. Effects of Microwave Radiating Equipments,* August 1959.

2. Livshits, N. N., "The Role of the Nervous System in Reactions to U.H.F. Electromagnetic Fields," *Biophys., 2,* 372 (1957).

3. Keplinger, M. L., "Review of the Work Conducted at the University of Miami," *Proc. of the Second Tri-Service Conf. on Biol. Effects of Microwave Energy,* July 1958.

4. Deichmann, W. B., Keplinger, M. L., and Lampe, K., *Effects of Microwave Radiation on Experimental Animals (24,000 Megacycles),* Department of Pharmacology, University of Miami School of Medicine, Coral Gables, Florida, sponsored by the Rome Air Development Center, Air Research and Development Command, USAF, 1959.

5. Livshits, N. N., "Conditioned Reflex Activity in Dogs under Local Influence of a V.H.F. Field Upon Certain Zones of the Cerebral Cortex," *Biophys., 2,* 198 (1957).

6. Von Euler, C., "Selective Responses to Thermal Stimulation of Mammalian Nerves," *Acta Physiol. Scand., 14,* Suppl. 45 (1947).

Biomedical Aspects of Microwave Irradiation of Mammals*

JOE W. HOWLAND, RODERICK A. E. THOMSON,
AND SOL M. MICHAELSON
Department of Radiation Biology
The University of Rochester School of Medicine
and Dentistry
Rochester, New York

THE MEDICAL DIVISION of the University of Rochester Atomic Energy Project is charged with the task of studying the biomedical aspects of 2800 mc pulsed microwave irradiation of the mammal. For this purpose an AN/MPS-14 Radar Search Unit at the Verona Test Site of Griffiss Air Force Base is used. In all exposures, the animal is housed in a Plexiglas cage and maintained in an anechoic chamber.

The details and physical characteristics of the exposures have been described previously (1). For the purpose of orientation, the exposure cage is shown in Figure 1. The usual thermal response of dogs exposed to 165 mw/cm² 2800 mc pulsed microwaves is illustrated in Figure 2.

Four phases of these studies include (a) Leucocyte Changes in Dogs After Exposure. (b) The Effect of Partial Body Exposure, (c) Influence on Body Water, and (d) The Effect of Intermittent Exposure.

LEUCOCYTE CHANGES IN DOGS AFTER EXPOSURE

Alterations in leucocyte levels have been reported to occur after exposure of the body to heat (2) and diathermy (3). It is of interest,

* Based on work done for the United States Air Force under Contract #AF30-(602)-1813.

Fig. 1. Exposure cage.

therefore, to characterize the changes in leucocyte levels after exposure to microwaves.

Materials and Methods

Mongrel dogs of either sex were exposed to 2800 mc, 12.5 cm, pulsed microwaves from the AN/MPS-14 radar unit at Griffiss Air

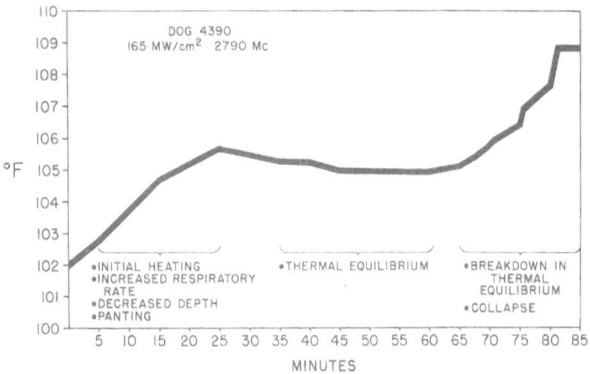

Fig. 2. Response of dog to microwave exposure.

Force Base or a 200 mc (1.5 m) continuous wave generating helical antenna at the University of Buffalo. The use of anesthesia, tranquilization, or sedation was avoided in order to observe a more valid physiologic response.

Power densities of 100 and 165 mw/cm² for different durations of time were utilized.

Blood samples were obtained by clean, single stab jugular puncture within 30 sec of exposure and examined by standard hematologic methods.

Results

Examination of the white cell changes indicate specific sensitivities related to frequency, field intensity, and/or duration of exposure. Total WBC changes reflect the sensitivity of the component cells (Table I).

At 2800 mc pulsed, 100 mw/cm² for 6 hr duration with mean rectal temperature increase of 1.8°F and minimal hematocrit change, there is an immediate polymorphonuclear leucocytosis with lympho-

TABLE I

Mean Leucocyte Values for Dogs Exposed to Microwaves

mc	Field intensity mw/cm²	Exposure (min)	No. of dogs	Sample	Temp.	Leuco-cytes per cmm	Neutro-philes per cmm	Lympho-cytes per cmm	Eosino-philes per cmm
2800	100	(330–360) 346	9 9 8	Control Post 1 day P̄	+1.8	13,540 13,770 16,630	9,159 11,802 11,930	3,435 1,385 3,242	639 208 795
2800	165	(90–120) 118	11 11 6	Control Post 1 day P̄	+3.0	14,120 11,240 14,400	9,380 8,323 9,236	3,348 2,053 4,390	730 468 588
2800	165	180	8 8 4	Control Post 1 day P̄	+4.4	12,360 14,940 17,840	8,520 11,520 14,065	2,896 1,982 2,853	648 659 332
200	165	(140–480) 325	4 4 2	Control Post 1 day P̄	+3.1	9,900 14,310 17,180	6,417 11,818 11,840	2,482 1,698 4,703	808 267 440

cytopenia and eosinopenia. Total leucocytosis is evident 24 hr post exposure. After 2 hr of exposure at 165 mw/cm² with 3°F rise in rectal temperature, there is an immediate depression in all the white cells with a definite hemoconcentration. Eosinopenia is still evident 24 hr later, while the other cells have returned to approximately normal levels. When the duration of exposure is increased to 3 hr with a resultant mean rectal temperature increase of 4.4°F and hemoconcentration, there is an immediate leucocytosis. The lymphocytes are depressed. Leucocytosis is more pronounced at 24 hr, and there is an eosinopenia.

At a field intensity of 165 mw/cm², 200 mc, there is an immediate total leucocytosis, polymorphonuclear leucocytosis with a lymphocytopenia and eosinopenia in the presence of hemoconcentration, and 3°F rise in rectal temperature. Total leucocytosis persists at 24 hr in the presence of hemodilution, and definite eosinopenia is still evident.

The most marked increase in hematocrit occurs when exposures of greater than 2 hr duration at 165 mw/cm², 2800 or 200 mc, are conducted (Table II). Hemoconcentration is not as pronounced after exposures at 2800 mc, 165 mw/cm² for less than 2 hr, or 100 mw/cm² for 6 hr.

Rats exposed to microwaves (2800 mc continuous) showed a marked increase in white blood cells 24 hr post exposure. Seven days later, the white cell count returned to approximately normal

TABLE II

Mean Hematocrit Values for Dogs Exposed to Microwaves

mc	Field intensity mw/cm²	Exposure (min)	No. of dogs	Temp. ΔT	Hematocrit		
					Initial	Post-exposure	
						Immediate	24 hours
2800	100	(330–360) 346	9	+1.8	47.9	49.5	47.6
2800	165	(90–120) 118	11	+3.0	47.4	50.0	47.1
2800	165	180	8	+4.4	45.8	53.6	46.0
200	165	(140–480) 325	3	+3.1	47.8	53.7	43.0

levels; and at 14 days, there was a secondary increase. By the 21st post exposure day, white blood cells of these animals had returned to the normal range.

Discussion

In general, prolonged exposure to microwaves produces a leucocytosis following the transient decrease in white blood cells seen after shorter periods of heating. Immediate post exposure leucocytosis, which is a reflection of polymorphonuclear neutrophile increase, is not correlated with the temperature elevation or hemoconcentration which develops from this type of exposure.

Leucocytosis is still evident 24 hr after prolonged microwave exposure, when the rectal temperature has returned to normal. The leucocytosis seen after prolonged microwave exposure is mainly a reflection of the increased polymorph levels. Exposures of long duration (ca. 6 hr) at lower power levels or longer wavelengths in which temperature increase is minimal, results in a marked eosinopenia and lymphocytopenia. This response is much smaller in exposures lasting 3 hr or less at higher power densities, which result in greater increase in rectal temperature. In most cases, a rebound of these cellular elements is apparent in 24 or 48 hr.

The decrease in lymphocytes and eosinophil levels seen after prolonged exposure resembles that reported to occur after slow continuous ACTH injection and may be indicative of hypothalamic or adrenal stimulation. Changes in urinary 17-hydroxycorticoids should be studied to confirm this hypothesis.

Recently, in dogs exposed several months previously to microwaves, a peculiar clumping of leucocytes resembling "leukergy" was observed (Fig. 3). According to Fleck (4), this condition occurs commonly in humans after various inflammatory situations such as rheumatism, arthritis, typhoid fever, and numerous other febrile conditions. Subsequent to this finding, similar clumping was noted during the summer months in some dogs with no previous history of exposure to any form of radiation. In these animals, the incidence or severity of the clumping is not nearly as great as that which is seen in dogs previously exposed to microwaves. Although it is difficult to assess the significance of this finding, it is quite possible that microwave exposure may exacerbate some latent condition in the animal making it more prominent after exposure.

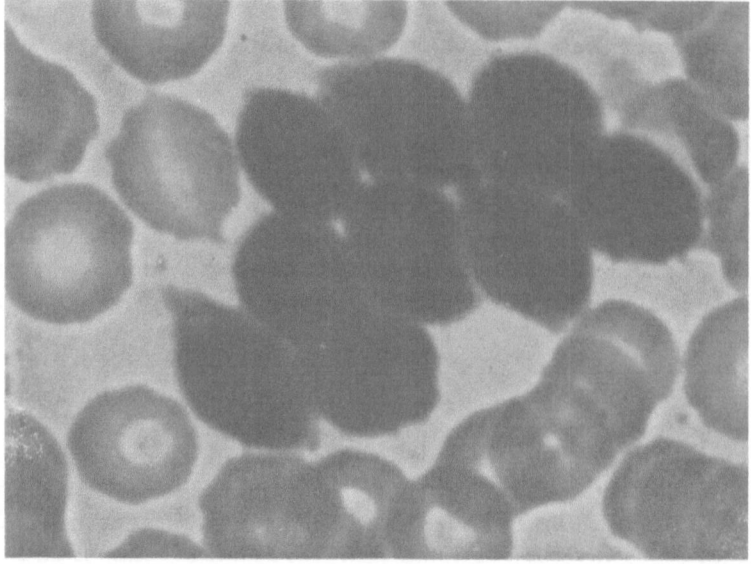

Fig. 3.

In addition, at certain times after exposure, an increased number of mitotic cells have been seen. This is rarely seen in normal dogs.

THE EFFECT OF PARTIAL BODY EXPOSURE

The response to exposure of limited portions of the animal's body to microwaves has been reported by this group and others (1, 5–8).

In order to characterize the clinical response of the animal in which a limited portion of the body is exposed, the following investigations were performed.

Materials and Methods

Dogs were exposed to 2800 mc pulsed or continuous microwaves at Griffiss Air Force Base and the University of Rochester. Head or trunk exposures were carried out on anesthetized animals. In all cases, the unexposed portion of the animal was shielded with absorbing material.

Results

Microwave irradiation of the head of the anesthetized dog results in a dramatic rectal temperature increase similar to that noted in whole body exposure. In the anesthetized animal, no accommodation to the initial temperature increase as noted in the normal dog is observed. Figure 4 is representative of a group of such experiments and shows that the rapid temperature reaction in head exposure to continuous wave exceeds that noted in whole body exposure to pulsed waves. The gradual continuous rise to critical temperature in 120 min or less is surprising (Fig. 5). Measurement of the skin temperature of a head-exposed animal shows a rise paralleling the rectal temperature.

To determine the relationship between the area exposed and possible contribution of central nervous system factors, an area of 500 cm^2 similar to the total surface area of the head was marked out on the side of a dog and exposed. The remainder of the animal was shielded. The response as shown in Figure 6 indicates a gradual rectal temperature increase requiring 220 min to reach critical levels. The rapid rise in axillary temperature reflects the direct radiation from the exposure site. The flank temperature obtained shows the effectiveness of the shielding.

To ascertain the effect of a smaller exposure area, a 188 cm^2 site over the rib cage was marked out and surrounded by shielding (Fig. 7). Critical levels were not reached in 270 min. Axillary and flank temperatures show the effectiveness of the shielding.

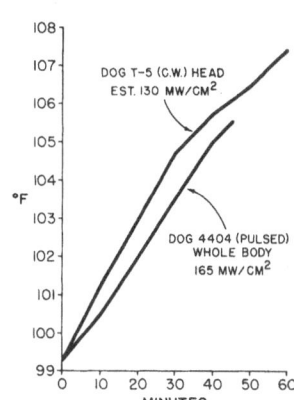

FIG. 4. Response of anesthetized dog to 2800 mc microwaves.

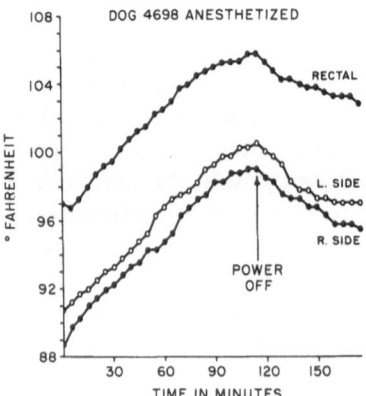

Fig. 5. Partial body exposure to 2800 mc microwaves (165 mw/cm²); surface area exposed (head) 500 cm².

Exposure of the head of the anesthetized dog with continuous wave invariably results in the production of burns. Marked swelling of the tongue also results, with production of numerous vesicles containing serous fluid localized largely at the base. Macroscopic examination shows involvement of skin, subcutaneous tissue, and muscles. A representative lesion is shown in Figure 8. In this particular case, the entire brain showed marked edema (Fig. 9), with a large area of hemorrhage immediately under the exposed site.

With head exposure to the pulsed 2880 generator, burns also occurred with edema and vesiculation around the base of the tongue.

Discussion

The evidence presented suggests that certain regions of the body show greater sensitivity to microwaves. Exposure of the head pro-

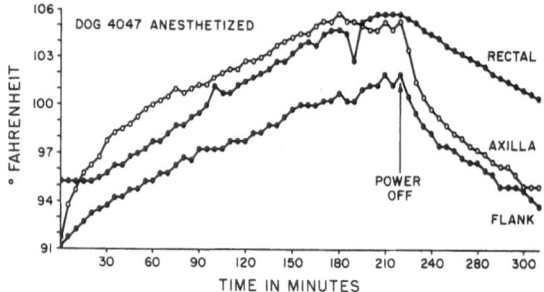

Fig. 6. Partial body exposure to 2800 mc microwaves (165 mw/cm²); surface area exposed (side) 500 cm².

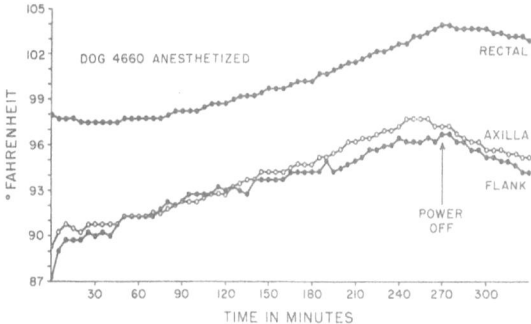

FIG. 7. Partial body exposure to 2800 mc microwaves (165 mw/cm²); surface area exposed (side) 188 cm².

duces a more dramatic reaction in extent and amount than that noted in areas of similar size elsewhere on the body. One should be interested in the possible contribution of the relative amount of circulating blood involved, the radiation of trapped heat into the cerebral cavity, the possible penetration of the radiation into the brain, and efficiency of the central nervous system in temperature exchange for a better solution to this problem. The best explanation at present

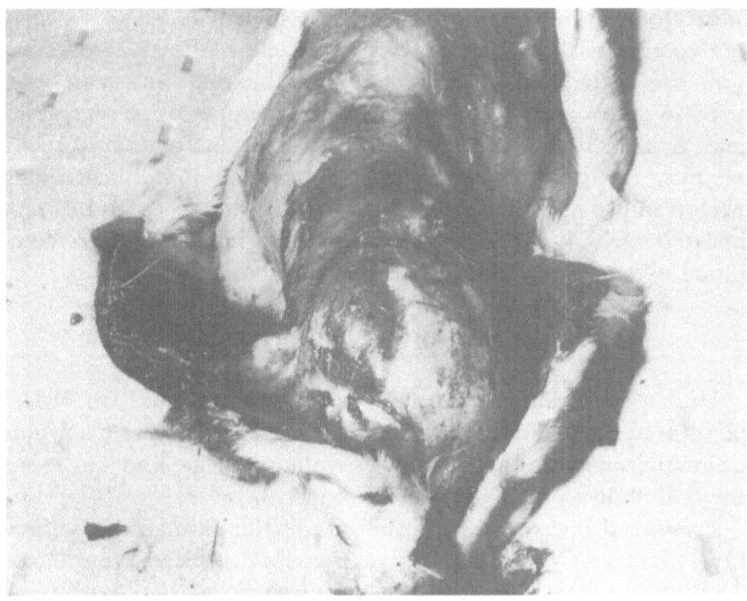

FIG. 8. Typical lesion.

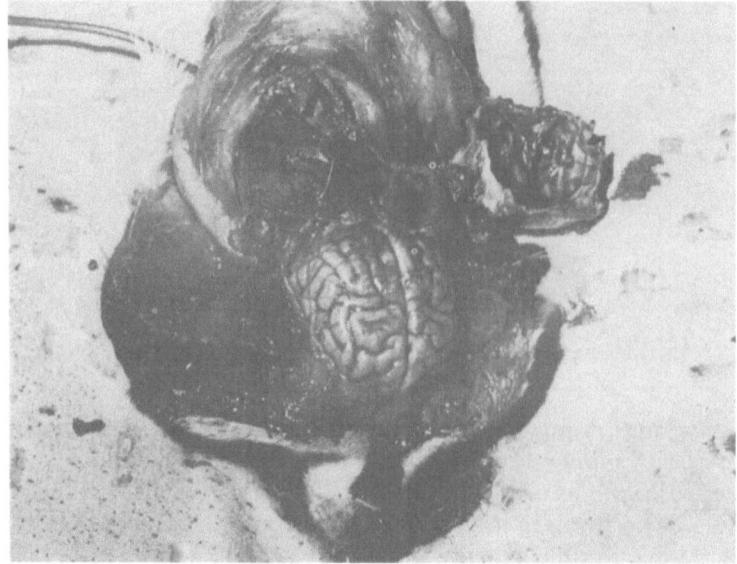

FIG. 9. Brain showing edema.

would include a consideration of local surface heating plus some contribution from brain or brain stem disturbance.

Exposure of local areas produces a reaction directly proportional to the size of the area exposed. It is probable that burns can be produced in small areas without generalized temperature reaction. If nerve tissue is involved (9) within the area, some temperature reaction may invariably be present. It is curious that burns did not develop in the anesthetized animal with partial body exposure but almost invariably were produced in the whole body-anesthetized animal exposed for similar periods of time.

INFLUENCE ON BODY WATER

Under conditions of thermal stress, the mammal, in order to maintain its normal temperature, will attempt to increase the volume of blood circulating through the peripheral areas, and at the same time, lose fluid to the external environment.

Prolonged microwave exposure places the animal in a condition of thermal stress. It is imperative, therefore, to investigate the part which water plays in the response to microwaves.

In general, change in body weight corrected for solids lost and food burned can be used to measure water loss (10). A correlation of body weight change with hematocrit alteration should provide an insight into the physiologic response of the animal attempting to maintain his body temperature while being exposed to microwaves. Of prime importance is the determination of the extent to which fluids can be restored during exposures of this type, with particular attention to abnormal or pathological sequelae. In order to determine whether these shifts could be better demonstrated in animals with chronic vascular defect, old dogs and certain dogs that had survived large doses of ionizing radiation several months to years previously were included. Representative studies will be demonstrated.

Materials and Methods

Adult normal mongrel dogs of either sex were exposed to 100 or 165 mw/cm^2, 2800 mc pulsed microwaves. For comparative purposes, additional animals were maintained in an ambient temperature of 103.5–106°F for similar periods of time in the same physical environment. Animals received water ad lib and were fasted 18–24 hr prior to exposure. Careful study of dogs exposed to microwaves indicated that no differences in reaction occurred in similar animals receiving controlled amounts of food and water, and those fasted with water as desired. Dogs that had received ionizing radiation in the LD 50/30 to LD 80/30 range several months or years previously were concurrently exposed to microwaves with the normal dogs by utilizing the double cage previously described (1).

In order to assess the part which hydration plays in the physiologic response of the animal exposed to microwaves, pairs of dogs were exposed simultaneously to 165 mw/cm^2. One animal was permitted water as desired, while the other dog remained unhydrated. In each trial, the animals were continuously exposed until one of the pair reached a temperature of 106°F before access to water was permitted. Approximately 200 ml aliquots of water at 50°F were delivered through Tygon tubing into a Plexiglas cup positioned in a corner of the cage. At all times, approximately 100 ml of water was left in the cup. To improve palatability, water was siphoned off and replaced as it became warm.

At arbitrarily chosen times during and at the termination of the exposure, each animal was accurately weighed within 10 g, and a

sample of blood was obtained for hematocrit determination. These
were then compared with pre-exposure values.

Results

Dogs exposed to 165 mw/cm² lose approximately 2% of their
body weight per hour (Fig. 10). At 100 mw/cm², body weight loss
averages 1.1%/hr. It is of interest to note that this weight loss is
linear after the first 30 min.

Maintenance of dogs at an environmental temperature of 103.5–
106°F for comparable periods of time results in a 0.6% body weight
loss per hour.

The interrelationship of hematocrit change, rectal temperature
increase, and weight loss between the normal dog and survivor is
illustrated in Figure 11.

It can be seen that the rectal temperature in the survivor increases
much more rapidly than in the normal dog, although the duration
of equilibration is approximately the same. Weight depression in
both cases is constant and linear. Hematocrit change in the normal
dog appears to be biphasic. Shortly after commencing the exposure,
hemodilution occurs. Approximately 1½ hr after beginning of the
exposure, hemoconcentration is evident. A second attempt at hemodilution is noted at approximately 2 hr followed by a final progressive marked hemoconcentration. In the survivor there is initial
hemodilution, followed by a progressively developing hemoconcentration with no apparent secondary hemodilution. It may further
be noted that both the survivor and normal dog lose the ability to

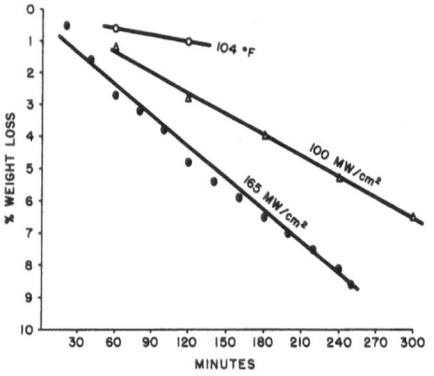

FIG. 10. Effect of 2800 mc microwaves and increased room temperature on body weight of normal dogs.

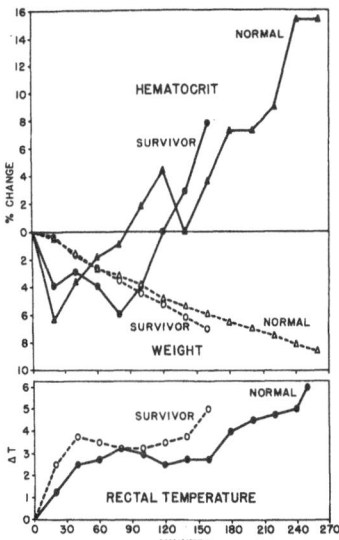

Fig. 11. Physiologic response in dogs exposed to 2800 mc microwaves (165 mw/cm²).

hemodilute at the time that thermal equilibration is starting to break down.

The normal dog is unable to maintain temperature or hematocrit stability at the point where he has lost 5% of his body weight. The survivor appears to lose this capacity to adapt at approximately 3% loss in body weight.

Weight, temperature, and hematocrit changes at the conclusion of exposure to 165 mw/cm² or increased environmental temperature are illustrated in Table III. It can readily be seen that exposure to 165 mw/cm² results in a body weight loss which is directly related to the duration of exposure. Hematocrit and temperature changes, although appearing variable, are not inconsistent. The greater hematocrit increase during the shorter exposure is coincidental with the first peak in the biphasic hematocrit curve.

Animals exposed to 100 mw/cm² show a smaller weight loss and temperature increase than do those exposed to 165 mw/cm². At this power level hemodilution occurs in contrast to the hemoconcentration evident at 165 mw/cm².

Dogs maintained at 103.5–106°F up to 6 hr display a minimal temperature increase, moderate weight loss, and variable hematocrit change.

Survivors (Table IV) exposed to 165 mw/cm² show a much

TABLE III

Response of Normal Dogs Exposed to 2800 mc Microwaves or Increased Room Temperature

Exposure	Duration of exposure (min)	Initial weight (kg)	Terminal weight (kg)	Difference (gm)	Difference (%)	Pre	Post	Difference	Temp. change (°F)	Remarks
165 mw/cm²	190	10.00	9.55	−450	−4.5	51.5	51.5	0	+4.50	Burn
	220	12.50	11.75	−750	−6.0	49.0	57.0	+8.0	*	
	270	11.20	10.40	−800	−7.1	46.5	51.0	+4.5	+3.25	Burn
	290	10.80	9.90	−900	−8.3	53.5	57.0	+3.5	+5.75	Burn
100 mw/cm²	60	16.10	15.90	−200	−1.2	58.0	57.0	−1.0	+2.00	
	360	8.25	7.80	−450	−5.5	46.5	38.0	−7.5	+1.50	
103.5–106°F	360	9.80	9.45	−350	−3.6	50.0	47.5	−2.5	+1.00	
	360	8.38	8.05	−330	−3.9	51.0	56.0	+5.0	+1.00	

*Thermistor probe defective.

higher rectal temperature increase in a shorter exposure time than do normal controls. Weight loss, although of a lesser degree, is comparable to that of the normal dogs if the shorter exposure time is taken into consideration. Hematocrit increase is greater than that of normal dogs per unit time and reflects an inability to hemodilute at these exposure times.

Survivors maintained at an increased environmental temperature display weight, temperature, and hematocrit changes comparable to normal dogs.

The response to hydration, starting at the time of critical temperature increase, is shown in Figure 12. Immediately following the intake of water, temperature falls rapidly to control levels. It has not been possible to carry hydrated animals to terminal critical temperatures in the 6 hr maximum experimental period. It must be noted that almost all animals exposed to these longer experimental times at 165 mw/cm² develop burns over the rib cage which are more extensive than noted in the nonhydrated group.

Table V shows the variation in the physiologic changes occurring in both normal and survivor groups. Averages include 3 animals per group. Exposures were in pairs as previously described. Length of exposure is largely determined by clinical condition.

TABLE IV

Response of Survivor Dogs Exposed to 2800 mc Microwaves
or Increased Room Temperature

Exposure	Duration of exposure (min)	Weight changes					Hematocrit changes			Temp. change (°F)	Remarks
		Initial weight (kg)	Terminal weight (kg)	Difference			Pre	Post	Difference		
				(gm)	(%)						
165 mw/cm²	65	13.85	13.65	−200	−1.4		54.0	55.5	+1.5	+6.50	Burn
	145	13.25	12.80	−450	−3.4		54.0	57.0	+3.0	+4.50	Burn
	160	11.45	10.65	−800	−7.0		51.0	55.0	+4.0	+5.00	—
	175	13.85	13.25	−600	−4.3		45.0	57.0	+12.0	+5.25	Burn
100 mw/cm²	60	12.25	12.10	−150	−1.2		47.0	49.0	+2.0	+2.25	
	335	14.10	13.30	−800	−5.7		53.5	53.0	*	+5.75	
	360	8.66	8.40	−260	−3.0		52.0	47.0	−5.0	+2.25	
103.5–106°F	300	14.72	14.14	−580	−3.9		51.0	45.5	−5.5	−0.40	
	300	16.74	16.03	−710	−4.2		53.0	52.5	−0.5	+2.25	

* Hemolysis.

In these highly variable data, certain observations are pertinent. In normal animals, hydration permits an extension of exposure to the limits of the experimental time. Hematocrit increase does not occur. A gain in weight results from the consumption of water as contrasted with a loss for the control group. A portion of this added water may exist in edema fluid lost into areas of early burn induction.

In survivors with probably impaired circulatory mechanisms, hydration does not effect a prolongation of permissible exposure (Table VI). Intake of water is a direct reflection of exposure time.

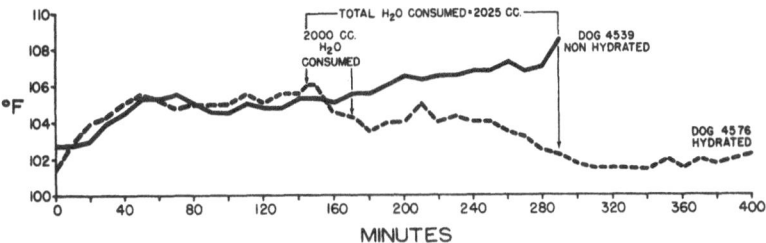

FIG. 12. Effect of hydration on rectal temperature response in dogs exposed to 2800 mc microwaves (165 mw/cm²).

TABLE V

Physiologic Changes in Dogs Exposed to Microwaves

2800 mc—165 mw/cm²

		Duration of exposure (min)	Water intake (ml)	Rectal temperature (°F)	Weight change (%)	Hematocrit
Non-hydrated						
	Normal	250* (190-290)	—	+4.50 (3.25-5.75)	−6.5 (4.5-8.3)	+4.0 (0-8.0)
	Survivor	128 (65-175)	—	+5.40 (4.50-6.50)	−3.1 (1.5-4.3)	+5.50 (1.5-12.0)
Hydrated						
	Normal	383 (300-460)	2422 (2025-3125) (200 ml/kg)	+2.50 (1.25-4.00)	+5.5 (3.3-9.2)	+4.8 (−1.5, −1.0, +17.0**)
	Survivor	148 (55-200)	1408 (950-2300) (100 ml/kg)	+5.00 (4.00-6.25)	+3.7 (2.5-4.9)	+5.3 (1.0, 2.0, 13.0)

* mean (range)   ** severe burn

Weight gain is similar to that in normal animals receiving twice the amount of water. Hematocrit and temperature increases are similar between the two groups. A high degree of variability occurs which bears a closer relation to exposure time than to other factors.

Discussion

An early manifestation of acute heat stress for the mammal is hemodilution, which occurs during the first 30 min of exposure and before body temperature increases. Upon prolonged exposure, hemodilution is reversed as a result of dehydration and hemoconcentration occurs. This early hemodilution is no doubt due to an influx of extravascular fluids as a result of the extensive peripheral vasodilatation which has taken place (11).

The dog exposed to 165 mw/cm², 2800 mc pulsed microwaves reveals a similar physiologic response.

The mean weight loss for dogs exposed to 165 mw/cm^2 equals 1.9%/hr. If exposure is continued until 5–7% of the body weight is lost, hemoconcentration is marked and thermal breakdown occurs.

Exposure to 100 mw/cm^2 results in 1.1% weight loss for each hour of exposure. When 165 mw/cm^2 is compared with 100 mw/cm^2, ratios of 1:1.65 for power density and 1:1.72 for body weight loss are derived. This may be indicative of a direct relationship of power density to insensible water loss, but requires additional experimentation at a third power level (preferably lower) to confirm this speculation.

If one were to use these ratios, assume no threshold for microwave effect, and add no exogenous water or food, calculation indicates that it would require approximately 40 hr of continuous exposure to 10 mw/cm^2, 2800 mc microwaves, uniformly distributed to reach a critical body temperature (106–107°F).

It is generally accepted that of the total body weight, 65% consists of extravascular fluid and 5% is intravascular (plasma). Compensatory fluid shifts occur in the mammal whenever body temperature is increased. Variations in blood volume may be detected by means of the hematocrit (12).

More or less as speculation, the hematocrit and body weight changes were utilized as indices of fluid alteration.

TABLE VI

Response of the Dog Hydrated During Microwave Exposure

2800 mc—165 mw/cm^2

Group	Dog	Duration of exposure (min)	Weight changes			Hematocrit changes			Temp. change (°F)	Remarks
			Initial weight (kg)	Terminal weight (kg)	Difference (gm)	Pre	Post	Difference		
Normal	4584	300	12.00	12.40	+400	51.5	50.0	−1.5	+2.25	Burn
	4413	390	12.90	13.40	+500	52.0	69.0	+17.0	+4.00	Very severe burn
	4576	460	10.90	11.00	+100	44.0	43.0	−1.0	+1.25	Burn
Survivors	4233	55	12.25	12.85	+600	51.0	53.0	+2.0	+6.25	
	A-27	190	9.85	10.10	+250	54.0	55.0	+1.0	+4.00	Burn
	4242	200	16.60	17.20	+600	57.0	70.0	+13.0	+4.75	Burn

Exposure of the dog to 165 mw/cm² until critical temperature (106–107°F) is reached results in a body weight loss equivalent to 10% of extravascular fluid. The survivor similarly exposed loses body weight equivalent to 5% of extravascular fluid. In the nonhydrated dog exposed to microwaves, fluid appears to move from the intravascular to extravascular spaces.

Upon hydration of the normal dog, fluid shift from the extravascular to intravascular compartment occurs. In the single case where the shift was reversed, marked edema was evident along the sides of the animal. Subsequent to this, an extremely severe burn developed.

In general, among hydrated normal dogs, 12% (6–20%) or less of ingested water is retained and body weight increases by approximately 3%.

It is of interest to attempt to ascertain the mechanism of water loss in animals exposed to microwave irradiation.

It is to be noted that during exposure animals do not urinate and only very rarely do they defecate. In addition, the major portion of insensible fluid loss in the dog is accomplished through alveolar respiration, foot pad sweating contributing very little to the total insensible fluid loss.

Assuming that water comprises 70% of the body weight, and in the dog insensible water loss through routes other than alveolar respiration is negligible, the following calculations can be made.

Per Cent Body Water Expired/Hour of Exposure

Normal	165 mw/cm²	2.3%
	100 mw/cm²	1.6%
	103.5–106°F	0.9%
Hydrated	165 mw/cm²	4.0%
Survivors	165 mw/cm²	2.4%
	100 mw/cm²	1.3%
	103.5–106°F	1.0%
Hydrated	165 mw/cm²	4.1%

These indicate that the water loss through the lungs in the nonhydrated dog parallels the energy to which the animal is exposed. At 165 mw/cm², approximately 1.5 times more water is lost than at 100 mw/cm² and 2.5 times more than at an increased environmental temperature. On hydration, the respiratory excretion of water is increased 1.7 times the above values.

The water loss in the survivor parallels the normal in rate per hour, but shortening of exposure times in all instances permits intake and excretion of only 50% of the amounts utilized by controls.

It must be mentioned that the contribution of the ambient temperature is critical in these experiments, inasmuch as the total water loss in warm expired air is greatly accelerated (13). Preliminary experiments with microwave exposure at an ambient temperature in the 103–106°F range show an active synergism of effect, reducing the tolerated critical exposure to less than 90 min at 100 mw/cm^2; ordinarily an innocuous level to 6 hr of exposure.

The clinical picture resulting from heating of animal or man would suggest that awareness of exposure to radiant or microwave heating is relative to the ambient temperature. Personal exposure demonstrates an immediate and acute perception of microwave energy at normal room temperatures. At levels exceeding normal body temperature perception appears to decrease. If such synergism exists and microwave damage is entirely or appreciably related to induction of thermal energy, consideration of the ambient energy should be made in the development of safe exposure values.

THE EFFECT OF INTERMITTENT EXPOSURE

The possibility of a cumulative effect of microwave exposure has confronted investigators for many years. Intermittent exposure to microwaves at fixed short time intervals should permit evaluation of the heating and cooling capability of the exposed subject. If this ability is related to a vascular and/or nervous mechanism which is altered with injury or age, the above technique could have great value in physiological measurement.

Materials and Methods

In a pilot study, 8 dogs were exposed in pairs, a normal and survivor of the same sex and approximate age. The field intensity used in all experiments was 165 mw/cm^2. When a rectal temperature of 107°F was reached by either of the animals exposed, the microwave field was interrupted for a regular arbitrary period of 20 min duration. Thus, after a cooling period of 20 min, the exposure was resumed for an additional 20 min followed by subsequent 20-min cooling and heating periods. Rectal temperatures were recorded

continuously. Body weight was recorded initially and at the conclusion of the exposure. Hematocrits were determined for each period
as shown.

Results

The results of these interrupted exposures are illustrated in Figure 13.

Analysis of the data indicates that at the end of the first heating
phase, survivors display a greater rectal temperature and hemoconcentration increase than do normal controls. During subsequent
heating phases, the temperature increase in the survivor is greater
than in the normal animal. The temperature at the termination of
the last 20-min period is markedly elevated among survivors in comparison with the normal control simultaneously exposed. Survivors
generally show a lesser degree of cooling than do normal dogs toward
the end of the terminal cooling phases.

For each degree rise in temperature, the hematocrit increases
1.6% among survivors and 2.5% among normal dogs. During the
cooling phase, for each degree drop in rectal temperature, there is a
1.6% decrease in hematocrit among survivors and 2.2% decrease in
normal dogs. This indicates a surprising lability of the vascular
mechanism in the normal dog.

The increased hemoconcentration in the survivor is related to the
increased body temperature reached during the heating phase and
the lesser reduction during subsequent cooling.

At the conclusion of cycling, body weight was recorded and com-

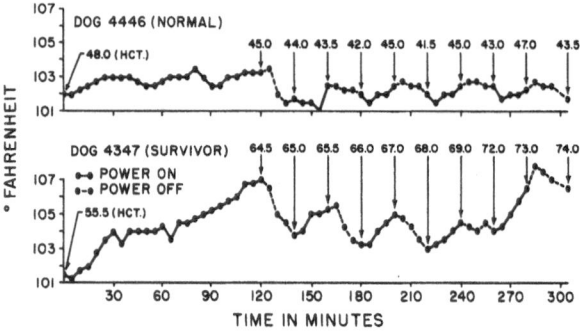

FIG. 13. Response to cycled 2800 mc microwaves (165 mw/cm²;
W. B.)

pared with the pre-exposure value. Two of the survivors showed a weight loss similar to their control normal dogs; two lost 30 to 40% less weight than their controls.

All of the survivors and 3 of 4 normal dogs developed burns which were evident at the end of the exposure period. No deaths occurred among the normal dogs. Two of the 4 survivors, whose weight reduction paralleled that of normal animals, died shortly after termination of the exposure.

Discussion

Intermittent exposure of the normal nonhydrated animal can be tolerated for extensive periods of time if breakdown of the thermo-regulatory apparatus is to be used as the criterion. The development of rib cage burns, however, suggests a definite relationship to the total exposure time whether the radiation is continuous or inter-rupted. From the data presented in the hydration experiments, it is doubtful that under tolerable operating conditions a critical temper-ature could ever be reached by intermittent exposure if water is taken as desired. However, the longer exposure time resulting would undoubtedly enhance the extent and severity of the burns produced. The capacity for cooling of specific local areas is markedly less than that of the whole body as measured by rectal temperature. During the early exposure periods, clinical edema of the rib cage can be noted which restricts the circulation to the area, thus allowing for heat ac-cumulation to critical temperature. These contributing factors plus the tissue injury from the irradiation produce the typical burns of microwave exposure.

As in the previously presented experiments, survivors intermit-tently exposed again demonstrate an abnormality of the vascular system, resulting in defective thermo-regulation. While experiments are in progress to determine whether this is a localized as well as a whole animal phenomenon, sufficient information has not been accumulated for definitive interpretation.

The acute or early effects of ionizing radiation have been actively investigated for many years. Life-span shortening, genetic damage, cataract induction, and increased incidence of neoplasia are being investigated for documentation of late manifestations of radiation injury. To date, very little attention has been paid to altered func-tional capability of the animal previously exposed to ionizing radiation.

Defective thermal regulation in the early period after whole body exposure to ionizing radiation has been described by Ryzhov (14), Hempelmann (15), and Hansen et al. (16). Defective thermal regulation as a late manifestation of radiation injury has not been adequately documented.

By the use of microwave exposure subsequent to ionizing radiation, marked residual effects of the latter can be seen. In these cases, a dramatic susceptibility exists to the induced hyperthermia and there is altered hemodynamic response, lethality, and incidence of localized burns.

It is apparent that among survivors there is a lowered threshold to the stress of microwaves and a defective thermal-regulatory mechanism. The observed defects are best explained by alterations in the vascular and/or nervous system which are most likely to occur in animals previously exposed to large doses of ionizing radiation.

Survivors from this colony maintained at an ambient temperature of 100°F for 10 days by Dr. Alan Keller at the U.S. Army Medical Research Laboratory, Fort Knox, Kentucky, and 105°F up to 6 hr at the Verona Test Site responded similarly to normal control dogs.

The difference in response of X-ray survivors exposed to both microwaves and increased environmental temperature may represent some physiologic relationship between microwaves and ionizing radiation which has not yet been determined. These differences suggest the possibility of a nonthermal effect of microwave exposure.

General Conclusions

A review of the accomplishments of the past year reveals some very intriguing and provocative manifestations of microwave exposure and provide approaches for future study.

In the area of fundamental research, the uniform thermal response and leucocytic alteration provide sufficient evidence to consider microwaves as a physical stressing agent. The advantages of this form of stress over other materials or agents is the fact that quantitation is possible to a considerable extent, and the animal, such as the dog, responds consistently and uniformly to this insult. Microwave exposure can provide a better insight into the physiology of thermal regulation, water balance, and the alterations in homeostatic capability caused by aging. Another aspect of these investigations is the possible interaction of microwave and ionizing radiation energy in the mammal.

In the realm of more practical aspects of microwave exposure, there is some evidence that at certain levels of exposure to microwaves, there may be direct brain stem effect.

The consumption of water during exposure to microwaves will depress the thermal response. However, if exposure is prolonged, burns in specific areas of the body may be produced.

A very important and serious consideration in investigations of the biological effects of microwaves is the determination of the maximal permissible exposure. Although 10 mw/cm^2 is apparently universally accepted as a permissible tolerance level, the lack of a time factor is somewhat disturbing and should be investigated. To date, there is no practical and reliable estimate of this exposure time. On the basis of calculations, it appears that in continuous exposure to 10 mw/cm^2, 2800 mc microwaves uniformly distributed, critical body temperature in the dog will not be reached in less than 40 hr. This estimate is made without consideration of a threshold for microwaves, which possibly does exist in the absence of exogenous water or food and lack of adaptive mechanisms. It can readily be seen that this duration of exposure under the conditions defined is quite improbable for man and therefore assures a sense of security. It should be pointed out, however, that power density/time tolerance factors do not take into consideration the possible occurrence of obscure or latent effects which are not readily demonstrable by the experimental or clinical methods available. The contribution of the environmental temperature in designating a safe tolerance level exposure time cannot be ignored. Preliminary experiments utilizing microwave exposure at increased ambient temperature show a definite synergism of thermal effect reducing the tolerance to microwaves.

These results and conjectures give impetus to the designing of experiments to provide a safe exposure level which would include, in addition to power density, a time factor and a consideration of the environmental temperature.

References

1. Howland, J. W., and Michaelson, S., "Studies on the Biological Effects of Microwave Irradiation of the Dog and Rabbit," ASTIA Document No. AD 212110, April 1959.

2. Forwell, G. B., "The Response of the Leucocyte Count in Man to Environmental Heat and Exercise," *J. Physiol., London, 124,* 66 (1954).

3. Bierman, W., "The Effect of Hyperpyrexia Induced by Radiation upon the Leucocyte Count," *Am. J. Med. Sci., 187,* 545 (1934).

4. Fleck, I., and Lille-Szyszkowicz, I., "Leukergie et leucoagglutination," *Sang, 27,* 589 (1956).

5. Searle, G. W., Imig, C. J., and Dahlen, R. W., "Studies with 2450 MC-CW Exposures to the Heads of Dogs," *Proc. 3rd Ann. Tri-Service Conf. Biological Effects of Microwave Radiating Equipments,* 54, August 1959.

6. Bach, S. A., Baldwin, M., and Lewis, S., "Some Effects of Ultra-high Frequency Energy on Primate Cerebral Activity," *Proc. 3rd Ann. Tri-Service Conf. Biological Effects of Microwave Radiating Equipments,* 82, August 1959.

7. Michaelson, S. M., Howland, J. W., Thomson, R. A. E., and Mermagen, H., "Comparison of Responses to 2800 mc and 200 mc Microwave on Increased Environmental Temperature," *Proc. 3rd Tri-Service Conf. Biological Effects of Microwave Radiating Equipments,* 161, August 1959.

8. Krusen, F. H., "Medical Applications of Microwave Diathermy: Laboratory and Clinical Studies," *Proc. Roy. Soc. Med., 43,* 641 (1950).

9. Nieset, R. T., Baus, R., McAfee, R. D., Fleming, J. D., and Pinneo, L. R., "The Neural Effects of Microwave Irradiation," RADC-TR-59-215, Nov. 1959.

10. Fulton, J. F., *Howell's Textbook of Physiology,* 15th ed., Saunders, Philadelphia, 1947, p. 955.

11. Bass, D. E., and Henschel, A., "Response of Fluid Compartments to Heat and Cold," *Physiol. Rev., 36,* 128 (1956).

12. Best, C. H., and Taylor, N. B., *The Physiological Basis of Medical Practice,* 4th ed., Williams and Wilkins, Baltimore, 1945, pp. 16, 19.

13. Lusk, G., *The Elements of the Science of Nutrition,* Saunders, Philadelphia, 1906.

14. Ryzhov, N. I., "Problem of Disturbances in Thermoregulation During Experimental Radiation Sickness of Dogs." Pathologic Physiology of Acute Radiation Sickness, P. D. Goringonton, ed., AEC-TR-3729, 265–271, 1958.

15. Hempelmann, L. H., Lisco, H., and Hoffman, J. G., "The Acute Radiation Syndrome: A Study of Nine Cases and a Review of the Problem," *Ann. Internal. Med., 36,* Pt. 1 (1952).

16. Hansen, C. L., Michaelson, S. M., and Howland, J. W., "The Febrile Response Following Upper Body X-Irradiation of Beagles," *Eighth Ann. Meeting Rad. Res. Soc.,* 1960.

Changes in the Ascorbic Acid Content in Lenses of Rabbit Eyes Exposed to Microwave Radiation

LORENZO O. MEROLA AND JIN H. KINOSHITA
Howe Laboratory of Ophthalmology and
Department of Biochemistry
Harvard Medical School and
Massachusetts Eye and Ear Infirmary
Boston, Massachusetts

FOR THE PAST YEAR the members of the Howe Laboratory of ophthalmology have been collaborating with Dr. Russell Carpenter and his associates at Tufts University to determine the biochemical changes in lenses of rabbits exposed to microwave irradiation. The irradiation of rabbit eyes and the classification of lenticular opacities were performed by Mrs. Claire Van Ummersen, Mr. David Biddle, and Dr. Russell Carpenter at Tufts University. In all these studies, microwave radiation with a frequency of 2450 mc, a wavelength of 12.3 cm, and a power density at 280 mw/cm² was used in a manner previously described by Dr. Carpenter and his colleagues (1). The length of exposure was varied depending on the severity of opacification desired.

There are two general aspects of the biochemical properties of the lens which can be studied to determine the cataractogenic effect under a given experimental condition. One is to examine the changes which occur to structural components and to normal constituents of the lens. The other is to follow changes in the dynamic aspects such as the carbohydrate metabolism and the energy producing mechanisms. In most instances these two general properties are interrelated. That is, if there were a change observed in the normal constituents of the lens, it would usually be the result of some interference in metabolism. Conversely, a decrease in metabolism would usually produce a change in the normal composition of the lens.

285

The effect of microwaves on the permeability of the lens was first studied. In our previous investigations we had established that changes in permeability of the lens alter the sodium and potassium content (2). The lens has a typical intracellular composition characterized by high potassium and low sodium content. An increase in permeability would be expected to produce a drop in K and an increase in Na as the lens equilibrates with ions of the intraocular fluids. For this reason, the lenses of eyes exposed to microwave radiation were examined for changes in K and Na levels. Since only one of the rabbit eyes was exposed to microwave rays, the lens from the other served as a control. Typical results obtained in these experiments are shown in Table I. It appears that levels of Na and K of irradiated lens were only altered when an obvious opacification had occurred. Where the opacities were minimal the cation content was essentially normal. The more severe the opacification, the greater was the shift in the cation distribution of the lens. A marked illustration of this is shown in the mature cataracts where the cation balance was grossly altered. In instances where transparency was maintained after irradiation, the levels of Na and K were normal. These results indicated that the factors which influence the cation distribution of the lens were not particularly sensitive to microwave irradiation.

A number of other constituents of lens were also equally unaffected. The levels of protein thiol groups, ammonia, glucose, and lactate fell only after an opacity had developed. These results sug-

TABLE I

The Effect of Microwave Radiation on the Cation Content of Rabbit Lens*

	Control	Severity of Opacities after Irradiation			
		Minimal	Moderate	Extensive	Mature
Na^+	15	15	17.8	35	124
K^+	119	110	110	101	37
$Na^+ + K^+$	134	126	128.8	136	161

* Values are expressed as meq/kg lens H_2O and are averages of 54 determinations. One of the eyes was unexposed and served as a control, while the other was exposed for 8 min to microwaves with a frequency of 2450 mc and a wavelength of 12.3 cm at a power density of 280 mw/cm^2. The rabbits were sacrificed 6 days after exposure, the eyes were removed, the lenses dissected, and the cation content was determined.

gested that the first sign of damage of lens by microwave radiation was the formation of opacities rather than any change in the chemical constituents.

This view was upheld until the levels of glutathione and ascorbic acid were examined. Glutathione occurs in lens in a higher concentration than in any other body tissue. Ascorbic acid is another reducing substance also found in high levels in this ocular tissue. The reason these two substances are found in such significant quantities and the role they play in the physiology of the lens are facts not as yet understood.

Ascorbic acid and glutathione in lens appeared to be more sensitive to microwave radiation than any of the other chemical constituents studied. However, the results of a large number of experiments conclusively showed that the drop in ascorbic acid occurred before any change in glutathione. Furthermore, the fall in ascorbic acid was observed before the development of opacification. The evidence for this is presented in Table II. The eyes of rabbits were exposed for 8 min and removed 18 hr after irradiation. The lenses of the microwave exposed eyes were transparent, had a normal content of glutathione, but had a substantially lowered level of ascorbic acid. The results seem to indicate that ascorbic acid is the most sensitive chemical constituent to be affected by microwave irradiation of the lens. This conclusion is noteworthy because in other forms of experimentally produced cataracts, such as in ionizing irradiation and diabetic cataracts, a decrease in glutathione content has been shown to be the first observable change. The distinguishing feature of microwave cataracts, then, is the fall in ascorbic acid content as the first sign of damage.

During the exposure period the possibility existed that an increase in temperature was the cause for the observed drop in ascorbic acid. This compound, being fairly unstable, could conceivably be affected by the increase in temperature incurred upon exposure of lens to microwaves. Experiments were, therefore, designed to check this possibility. The results (Table III) indicate that removal of the lenses ½ hr after the microwave exposure revealed that no change in the ascorbic acid content had occurred. This finding rules out the possibility that increase in temperature during exposure was the cause of the disappearance of ascorbic acid. The drop in the level of ascorbic acid was not observed 6 hr after irradiation, but was found in lenses removed 18 hr after exposure. This indicates that

Merola and Kinoshita

TABLE II

The Effect of Microwave Radiation on the Ascorbic Acid and Glutathione Contents of Rabbit Lens*

Rabbit no.	Ascorbic acid			Glutathione	
	mg %		% dec.	mg %	
	Control	Exp.		Control	Exp.
214	27.6	20.8	24	376	376
215	21.8	15.1	30	376	376
216	24.0	19.6	21	392	382
218	24.0	19.0	20	410	405
220	18.7	15.4	18	310	310
221	22.0	19.0	15	352	346
222	25.2	18.1	29	340	340
226	22.3	14.2	36	360	343
227	21.9	17.1	22	310	310
228	22.4	16.4	27	312	312
229	26.6	20.0	24	302	302
230	21.3	16.6	24	297	297
231	19.8	15.0	24	330	330
232	21.0	16.6	21	360	360
233	23.0	18.6	19	310	310
252	17.2	12.7	25	328	324
254	19.4	13.7	29	304	304
256	17.2	14.3	17	286	290
259	18.1	14.4	20	296	300

* Rabbits were sacrificed 18 hr after one of their eyes had been exposed for 8 min to microwave energy at a frequency of 2450 mc and a wavelength of 12.3 cm. The power density was 280 mw/cm². The lenses were removed from exposed and unexposed eyes and precipitated in trichloroacetic acid. The ascorbic acid content of the control and exposed lenses were determined on aliquots of protein-free filtrate by titration with 2,6-dichlorophenolindophenol. Glutathione was determined by the nitroprusside method on another aliquot of the same filtrate.

the decrease in ascorbic acid which results upon microwave irradiation of lens does not occur immediately after exposure but that it develops after a latent period of 6–18 hr.

Not too much is known about the presence of ascorbic acid in lens. It has not been clearly established whether this reducing agent is synthesized in lens. Most of the evidence is against its *de-novo* synthesis in this ocular tissue. It is generally accepted that ascorbic acid is actively secreted into the aqueous humor, and this accounts for its high level not only in this ocular fluid but also in the lens. We wondered whether the reason for the drop observed in lenses of rabbit eyes exposed to microwave radiation was due to a decrease of this substance in the aqueous humor. To investigate this possibility samples of the aqueous humor were withdrawn from the eyes exposed

TABLE III

**Time Course of Decrease in Lens Ascorbic Acid
after Microwave Irradiation***

Rabbit no.	Period after exposure, hr	Ascorbic acid, μg/lens	
		Control	Exposed
142	½	100	100
144		116	116
145		98	104
147		108	104
148		100	100
149		96	96
150		100	100
166		110	110
167		111	120
168		120	120
172		110	110
174		100	100
153	6	92	92
154		86	86
155		92	92
162		99	99
151	18	99	71
163		99	78
164		113	92
165		107	78
177		122	88
178		110	75
180		125	100
183		125	88
193		100	80
207		75	67
209		92	75
211		89	75

*Exposure to microwaves with a power density of 280 mw/cm² was for 8 min.

to microwaves and from the control. The ascorbic acid level was measured in the intraocular fluid as well as in the lens. The exposure was for 8 min and the eyes were removed 18 hr after irradiation. In all cases reported in Table IV, no opacification was observed in the lenses. The assays revealed that little or no change of ascorbic acid level occurred in the aqueous humor of the irradiated eyes. With the exception of rabbit #221, which showed a 15% drop in ascorbic acid of the aqueous humor, the other 14 cases revealed no significant drop. On the other hand, the ascorbate level was consistently lower in the irradiated lenses than that of the controls. These results indicate that when a drop in the ascorbic acid content of the lens occurred there was no corresponding decrease of this substance in the aqueous humor. Other reasons for the fall in ascorbic acid of lens exposed to microwaves and the possibility that this decrease influences the course of cataract formation are now being investigated.

TABLE IV

The Effect of Microwave Radiation on the Ascorbic Acid Level in Aqueous Humor and in Lens*

Rabbit no.	Aqueous humor, mg %			Lens, mg %		
	Control	Exp.	% change	Control	Exp.	% change
216	24	24	0	24.0	19.6	− 20
220	15.4	15.9	+ 3	18.7	15.4	− 18
221	19.7	16.8	− 15	21.7	18.6	− 15
226	25.4	25.4	0	22.3	14.2	− 36
238	20.0	19.8	0	17.8	15.8	− 11
240	20.6	20.0	0	13.7	12.7	− 8
256	21.4	21.0	− 2	17.2	14.3	− 17
254	21.8	20.0	− 8	19.4	13.7	− 29
252	14.8	13.9	− 6	17.2	12.7	− 25
259	21.8	21.0	− 4	18.1	14.4	− 20
269	23.8	24.8	+ 4	24	15.4	− 35
273	14.2	13.7	3	21	14.0	− 34
274	16.6	17.0	+ 3	21	17.2	− 22
280	25.2	26.2	+ 4	22	16.6	− 25
283	31.0	31.6	+ 2	18.8	14.4	− 23

* The eyes were exposed to microwave energy for 6 min at 280 mw/cm² power density and the rabbits were sacrificed 18 hr after exposure. Immediately a 0.1 ml sample of aqueous humor was withdrawn from the exposed and unexposed eyes and ascorbic acid level was determined. The ascorbic acid content in the lens was determined as described previously.

References

1. Carpenter, R. L., "Experimental Radiation Cataracts Induced by Microwave Radiation," *Proceedings of the Second Tri-Service Conference on Biological Effects of Microwave Energy,* July 1958, pp. 146–168.

2. Merola, L. O., Kern, H. L., and Kinoshita, J. H., "The Effect of Calcium on the Cations of Calf Lens," *Arch. Ophthalmol. (Chicago), 63,* 830 (1960).

Preliminary Results of Studies of the Lenticular Effects of Microwaves Among Exposed Personnel

Milton M. Zaret and Merril Eisenbud
Departments of Ophthalmology and Industrial Medicine
New York University Medical Center
New York, New York

INTRODUCTION

THIS STUDY IS THE OUTGROWTH of two previous series of investigations. First, the cataractogenic potentiality of microwave radiations in the laboratory animal under controlled environmental conditions was clearly demonstrated by Carpenter and others. Second, data obtained from examination of a group of employees at the Rome Air Development Center were suggestive of an increased incidence of abnormal lens findings among microwave workers. As a result, an ad hoc Eye Study Group was convened at the request of the Commander, Rome Air Development Center, for the purpose of reviewing the lens findings among the Rome employees. The study group concluded that the types of lens defects found at Rome were not different from those noted in normal, unexposed individuals, that such defects do not necessarily interfere with vision, and that the reported defects should be distinguished from cataracts, which represent a gross degree of lens opacification resulting in reduced visual acuity.

Herein lies one of the principal difficulties encountered in the design of our study. Many minute defects are spontaneously present in the lens. Some of these defects seem to remain stationary for life. Others are associated with progression of opacification and may ultimately result in cataract formation. When this occurs the rate of

293

progression is usually so slow that years pass before the process is complete.

It should be noted that ophthalmologists do not ordinarily diagnose an eye as having a cataract unless loss of visual acuity has occurred or the defects in the lens are massive as compared to the defects found in the Rome group. Moreover, ophthalmologists do not routinely study the lens in great detail for minute defects, and there is no standard method of conducting the slit-lamp examination or of documenting the morphological and geographical findings.

The Eye Study Group recommended that the original survey should be expanded to include additional military personnel and that similar studies should be carried out on selected employee groups from industrial facilities where microwave exposures exist, in order to obtain a larger sample and more meaningful statistics. The study group further suggested that, if possible, the slit-lamp examinations should be performed by one ophthalmologist to ensure uniform observations.

Anatomical Factors

At this point, it may be helpful to review some of the anatomical characteristics of the normal lens. The lens is entirely ectodermal in origin and at an early embryonic stage of development is isolated from the rest of the body by its capsule. Although this tissue is isolated from the cellular components of the body, it continues to grow throughout life without any new cells entering the lens or any lenticular cells leaving. Being avascular, the lens is at a disadvantage by not having as effective a cooling system as the other tissues of the body as well as not having available macrophages or replacement cells, as occurs in the reparative processes elsewhere in the body.

In the center of the lens is the fetal nucleus, surrounded by the adult nucleus which, in turn, is surrounded by the cortex. Lining the anterior surface of the lens is a one-cell-thick layer of epithelial cells which extends to the equatorial region, where they undergo transformation into lens fiber cells. The peripheral surface of this tissue is enclosed by the capsule.

Due to the optical qualities of the eye, the lens can be examined by a slit lamp, a type of biomicroscope having a light beam which can be altered in size and intensity. In addition, the light may be focused upon any level of the lens concurrently with the viewing microscope, so that only the portion brought into focus may be examined. By slit-lamp examination the various regions of the lens may be recognized and imperfections noted.

Injury to any portion of the lens may result in loss of transparency, a condition termed opacification. This may occur in localized regions of the lens or diffusely throughout the lens substance. When opacification has been established, it may remain stationary throughout life, it may progress, or, after some progression, it may become stationary.

Sporadically, cases of cataract formation in humans who have been exposed to microwave radiations have been reported. These cases have exhibited extensive, dense opacification which represents the end result of massive lens damage. Some of these patients have been followed closely by ophthalmologists who have attempted to determine whether casual relationship existed. In addition, the above-mentioned study made at the Rome Air Development Center was highly suggestive of damage to the posterior cortex layer of the lens, a region that may be affected early in some types of cataract formation.

Method of Study

Our study was designed with the following purposes in mind:

1. To establish a standardized method of examination of the crystalline lens of the eye.

2. To examine exposed and control personnel to determine the kinds of lenticular defects and the frequency with which they occur.

3. To gather information about microwave and ionizing radiation exposure of the subjects studied.

4. To determine if correlations exist between the ophthalmological findings and exposure data.

Preliminary studies were made to establish a standardized method of slit-lamp examination in association with the use of the Donaldson stereo-camera, a device which photographs the lens of the eye in stereopsis. By combining these two methods, it was possible to document many types of lens defects, especially those having potential clinical significance.

At the same time, environmental exposure history questionnaires and criteria for selecting exposed and control subjects were developed.

The original data reported by the Rome Eye Study Group were

examined, and it was found that an apparent linear relationship exists between the age of an individual and the number of lenticular defects of all types. This is shown in Figure 1, which gives the defects per eye in each of the age classes of the unexposed personnel examined. This served to emphasize the importance of aging on the frequency of lenticular defects.

It soon became apparent that every lens has some defects. It was decided that for the purpose of our study, only those defects would be recorded that could be verified readily by slit-lamp examination or documented by photography.

Following our initial survey at site A, a large military research establishment, it was decided to express the results of the eye examination according to the following nomenclature:

1. Minute Defects. This category included all defects such as granules, vacuoles, and tiny opacities which are ordinarily too small to be photographed or too numerous to be individually catalogued.

2. Opacification. Before becoming clinically cataractous, the lens may assume an irregularly diffuse cloudiness, or discrete regions of the lens may become markedly opacified. If such a region of the lens is close to the focal plane of the camera, it photographs well and is easily documented. It is desirable to have two exposures of each lens, one focused in the anterior cortex, and the other focused in the posterior cortex.

3. Luminescence. Optical luminosity occurs when the beam of light from the slit-lamp traverses the lens and the various regions may be distinguished from one another by the differing degree of haziness created by the light beam.

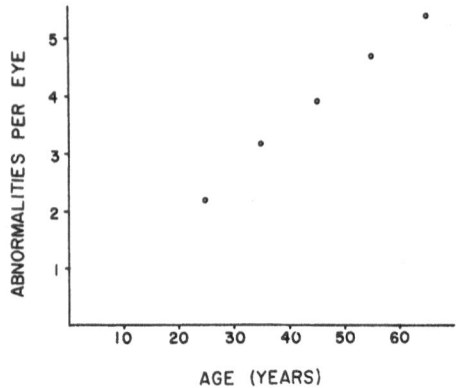

FIG. 1. Number of defects per eye per age class in unexposed personnel.

4. *Sutural Defects.* The sutures of the lens can usually be observed by slit-lamp examination. As this region of the lens is prominently involved in experimental microwave cataracts, it was mandatory to include this region of the lens as a category for investigation. Thickening, banding, and striations were noted as sutural defects.

5. *Posterior Polar Defects.* The posterior subcapsular cortex, especially in the polar region, is a frequent site in which early changes occur, regardless of the type of insult to the lens. In some cases of ionizing radiation injury it is pathognomonic for the doughnut-shaped subcapsular defects to progress to the stage of polar, subcapsular cataract.

The above categories of the eye examination were graded on a relative scale in the following manner: 0 for insignificant numbers or degree, 1 for small numbers or minor degree, 2 for moderate numbers or degree, and 3 for large numbers or major degree short of clinically recognized cataract.

The method of selecting subjects, like other aspects of this program, has evolved gradually. At site A the subjects were referred for examination by simply selecting a group who were known to have had the most direct and prolonged contact with microwave-producing operations. This practice was again followed at site B, an industrial installation. However, it soon became apparent that although a wide range of exposures was available in the groups being submitted for examination, there was no assurance that the subjects were a representative sample of the entire exposed population, or that the variations in severity of exposure would provide the necessary internal controls necessary to answer the question as to whether the ophthalmological findings are related to the extent of microwave exposure. It was therefore decided that, to the extent possible, controls should be selected on a one to one basis, and that they should age-match the exposed population and be selected from the same site as the exposed employees. Use of controls on a one to one ratio to exposed personnel seems to yield differences having a high degree of confidence. The technique of age-matching the controls and exposed personnel serves to eliminate a potential bias due to age, and selection of controls at the sites of observations reduces the need to deal with the unknown questions as to whether or not lens defects are affected by social, economic, and geographical considerations.

It was also decided that it would be desirable for all exposed employees at each installation to fill out exposure history questionnaires so that the sample selected for study would be representative of the entire exposed population.

The decision to use age-matched controls was made after the examinations of exposed personnel was completed at site B. That the subjects were unexposed to microwaves was thus known at the time of ophthalmological examination at this site. However, at site C, the controls and exposed subjects were intermixed and the examinations were conducted in such a way that the ophthalmological findings were recorded without any knowledge of the exposure history of the subject.

Evaluation of Environmental Factors

Insofar as assessment of environmental factors is concerned, retrospective evaluation of occupational history is always difficult in epidemiological investigations. In this study, the difficulties are even greater than those normally encountered because a great many factors influence the severity of exposure, and there is no known quantitative method by which these can be treated. The personnel included in this study have been exposed to microwaves for periods of time ranging from only a few days to almost 18 years. Many of the men have been exposed to a wide range of frequencies. Average power outputs have varied from less than a watt to several kilowatts. The work has been performed under all kinds of operating conditions, some of which may have given rise to ionizing radiation exposure.

It was immediately apparent that a quantitative reconstruction of past work history would not be feasible. On the other hand, it appeared reasonable to assume that if one could collect all the available occupational information from each employee, it might be possible to divide the exposed subjects into two or more classes, according to the severity of their exposure.

A questionnaire was designed for this purpose (see Appendix). On these pages the employee is asked to list the various jobs in which he was exposed to microwaves, the duration of known exposure, and the kind of employment. In addition, information is requested about types of equipment, the average power outputs, their frequency, and the manner in which the power output was terminated. The employee is asked if he recalls having felt heat from either waveguides or antennae, his distance from the equipment, and many other questions.

The cumulative ionizing radiation exposure of the personnel cannot be estimated because only in a few cases were film badges worn at all, and then only for a limited period of time. However, some of the subjects, such as those who worked only around antennae, were not subject to ionizing radiation exposure, whereas those who worked in the immediate vicinity of large microwave-producing tubes were subjected to ionizing radiations as well as microwaves. This fact should make it possible for us to examine, at least qualitatively, the question of whether potential ionizing radiation exposure potentiates the effects of microwave exposure. Moreover, in the prospective studies which, it is hoped, will emerge from the present microwave investigation, it should be possible to deal more quantitatively with this factor inasmuch as all subjects who may be exposed to ionizing radiation in the course of their work will wear film badges.

For the purpose of the present pilot study semi-quantification of the exposure is undertaken by the scoring system tabulated in Table I. The exposure factors are graded from 1-3. These include power output, distance from tube or transmission line, a history of having looked into an energized waveguide, a history of having sensed heat from a waveguide, and a history of having sensed heat from the

TABLE I

Method of Calculating Microwave Exposure Index*

Factor	Weight		
	1	2	3
1. Power output (av. watts)	<100	100-1000	>1000
2. Distance from tube or transmission line (feet)	>10	<10	–
3. Looked into energized waveguide (no. of times)	1-10		>10
4. Felt heat from waveguide	<10	Hands only >10	Head or whole body >1
5. Antennae exposure (location and time)	Rear or sides	Front (seconds or minutes)	Front (hours)
6. Antennae heat (time)	Seconds	Minutes	$>$Hour
7. Antennae power (av. watts)	<100	100-1000	>1000

* Exposure Index $= (1 \times 2 \times 3 \times 4) + (5 \times 6 \times 7)$.

antennae.   The exposure index assigned to each employee is obtained
by taking the product of items 1 × 2 × 3 × 4 and adding this to the
product of 5 × 6 × 7.   By this system, it is possible for an individual
to attain a maximum score of 81.

It should be noted that this system is only a method by which an
attempt can be made to sort the exposed employees into exposure
classes of relative severity.   For the purpose of this pilot study, we
have used only three classes in the analysis.   Those having a score of
zero are our controls.   We call those having a score of less than 5 the
lightly exposed group, and consider those having a score of greater
than 5 to have been relatively heavily exposed.   This scoring system
will undoubtedly undergo modification in the future and, as the
population becomes larger, it may be possible to divide the group into
additional classes of exposure severity.

Description of the Examined Groups

The basic characteristics of the population examined at the three
sites are given in Table II.   The exposed men numbered 170.   Of
these, exposure histories have been completed on 146, and these are
divided, for the purpose of comparison, into 72 having an exposure
index less than 5, and 74 having an index greater than 5.   The mean
ages of all groups are comparable, but the average duration of exposure at site C is less (40 months) than at sites A (70 months) and
B (77 months).

TABLE II

Summary of Population Examined

Site	Controls		Exposed			Av.	
	Number examined	Mean age	Number examined	Exposure index*		Mean age	Av. exp. months
				# <5	# >5		
A	0	—	66	22	30	34	70
B	29	35	37	17	14	35	77
C	41	36	67	33	30	32	40
Total	70		170	72	74		

*Some exposure histories were incomplete.

Findings

No reduction of visual acuity due to cataracts was found at any site.

One patient was found to have intumescent lenses, a condition where the lens is swollen due to hydration. Questioning revealed that about 1 week prior to this eye examination he was inadvertently exposed to microwave energy, and it is possible that the intumescent lenses were due to an acute radiation injury. Environmental data are being reconstructed.

As noted above, the objective of this investigation is to record the types and frequencies of lenticular changes that occur among healthy males of various ages, and to determine if these changes are qualitatively or quantitatively different in groups of individuals who have been exposed to microwave radiation. Should such changes be noted, an attempt will be made to ascertain the relative importance of the many environmental factors that might be involved in the production of these changes. Some of these factors are concomitant exposure to ionizing radiation, length of employment with microwaves, wavelength, and power.

It is, of course, premature to draw conclusions from the relatively few examinations that have been completed to date, but there is some value in noting briefly the relationships that have emerged thus far on the basis of the limited statistical treatment that has been permissible.

Among the 70 controls examined, there was a definite tendency for the lenticular changes to increase with age. For example, no grade 2 or 3 opacifications were observed among the men aged 20 to 32 years, whereas among the group aged 33 to 55 years, 27% of the men were so scored. This is a highly significant difference, $(P < .01)$ when tested by the method of chi-square.

In order to avoid any bias due to differences in age of the control and exposed populations, comparisons of the frequencies of observed lenticular changes were made by selecting aged-matched pairs of control and exposed subjects. To test for differences between the two groups, the pairs were first classified as follows for each type of lenticular change: Class 0, pairs in which the control and exposed member were scored alike; Class $(-)$, pairs in which the exposed member received the lower score; Class $(+)$, pairs in which the exposed member received a higher score. The results of comparisons among 67 pairs is given in Table III.

These data indicate that among pairs showing a difference, the exposed members have a highly significant number of higher scores for opacification, luminescence, and posterior polar defects. Moreover, the tendency for the exposed member to have a higher total score is also highly significant.

There appears to be an association between opacification, luminescence, and posterior pole defects. (Not only is this strongly suggested by Table III, but also by the chi-square test which indicates a highly significant association.) No such association is demonstrable for minute defects nor for sutural defects.

The exposed personnel were divided into two classes having exposure scores of 1–4 and ≥ 5. Using the method of chi-square, an apparent association is demonstrable between opacity score and exposure index, but the relationship is not entirely conclusive.

Summary

1. A standardized method of examination of the crystalline lens of the eye and reporting of findings has been established.

2. A method of evaluating relative severity or exposure to microwave radiating equipment has been established.

3. No late lens defects were found to be peculiar to microwave exposure. All defects noted were of types observed in the normal population.

TABLE III

Comparisons of Age-matched Pairs*

Findings	Comparison of scores				P
	0	(−)	(+)	−/(+ and −)	
Minute defects	19	20	28	20/48	
Opacification	30	7	30	7/37	<.01
Luminescence	37	7	23	7/30	<.01
Sutural defects	39	13	15	13/28	
Posterior polar defects	26	5	36	5/41	<.01
Total defects	11	13	43	13/56	<.01

* Sites B and C.

4. Statistically significant increases in the occurrence of posterior polar defects, luminescence, and early opacification of the lens were found in the exposed group as compared to the control group of personnel.

5. The data suggest further that the frequency of defects may be dependent upon relative severity of exposure to microwaves.

6. The study is being continued and more definitive conclusions will become permissible as the size of the observed population is increased.

ACKNOWLEDGMENT

The authors gratefully acknowledge the assistance of the following: Herbert Schmidt, M.D., for his valued advice; Mr. Blaine Howard, who assisted in the design of the exposure history questionnaire; and Miss Lee Herrera, who undertook the statistical analysis of the data.

APPENDIX

NEW YORK UNIVERSITY MEDICAL CENTER
550 First Avenue   ·   New York 16, N.Y.
Institute of Industrial Medicine Environmental Radiation Laboratory

History of Work with Microwaves

YOU ARE BEING ASKED to fill out this questionnaire as part of a large study in which we are one of a number of participating industrial and military organizations.

We appreciate the fact that in some cases a great many years may have elapsed since you first began to work with microwaves and that it may be difficult for you to recollect all of the detailed information we have asked you to provide. All we request is that you be patient with this questionnaire and fill it out to the best of your ability. If there are any parts of it which are not clear to you, your supervisor will attempt to assist you.

Name _____   Age _____   Badge No. _____
      (Please print)

Address _____

_____

I. List only those places of employment in which you worked with radar or other microwave equipment. List present employer first and work backwards.

a. Employer's
Name_____

Employed from _____ to _____

Job Titles: 1. _____

2. _____

3. _____

Total Number of Months Employed in the Following Categories

a. Research and development of microwave components

_____

b. Microwave components assembly for production _____

c. Operation of radar or other microwave apparatus

_____

d. Installation, maintenance, and test of microwave apparatus

_____

e. Other   _____

Total _____

b. Employer's
Name_____

Employed from _____ to _____

Job Titles: 1. _____

2. _____

3. _____

a. Research and development of microwave components

_____

b. Microwave components assembly for production _____

c. Operation of radar or other microwave apparatus

_____

d. Installation, maintenance, and test of microwave apparatus

_____

e. Other   _____

Total _____

c. Employer's
Name_____

Employed from _____ to _____

a. Research and development of microwave components

_____

Job Titles: 1. _____

            2. _____

            3. _____

b. Microwave components assembly for production _____

c. Operation of radar or other microwave apparatus

_____

d. Installation, maintenance, and test of microwave apparatus

_____

e. Other   _____

       Total _____

d. Employer's Name _____

Employed from _____ to _____

Job Titles: 1. _____

            2. _____

            3. _____

a. Research and development of microwave components

_____

b. Microwave components assembly for production _____

c. Operation of radar or other microwave apparatus

_____

d. Installation, maintenance, and test of microwave apparatus

_____

e. Other   _____

       Total _____

II.

a. Did you ever wear a film badge?   Yes _____   No _____

b. If yes, on which job and during what period of time.

Places of Employment	From	To
a. _____	_____	_____
b. _____	_____	_____
c. _____	_____	_____
d. _____	_____	_____

III.

a. Did you ever work near a transmitter tube from which the shielding was removed while the high voltage was on?   Yes ___ No ___

b. If yes, fill out the following:

			Check Total Time Elapsed (approx.)			
Tube Type	*Aver. Power*	*Peak Voltage*	*Sec.*	*Min.*	*Hrs.*	*Longer*
___	___	___	___	___	___	___
___	___	___	___	___	___	___
___	___	___	___	___	___	___
___	___	___	___	___	___	___

IV. List the principal types of microwave generating equipment with which you have worked.

	a.	*b.*	*c.*	*d.*
Type of equipment	___	___	___	___
Average Power	___	___	___	___
Freq. or Band	___	___	___	___
Number of months	___	___	___	___
Date of first exposure	___	___	___	___
Power terminated (use check mark)				
Dummy load	___	___	___	___
Outside antenna	___	___	___	___
Within room	___	___	___	___
Distance from equipment				
Less than 10 ft.	___	___	___	___
10–20 ft.	___	___	___	___
Greater than 20 ft.	___	___	___	___
Your work was:				
a. research and development	___	___	___	___

b. assembly of microwave
components _____ _____ _____ _____

c. operation of microwave
equipment _____ _____ _____ _____

d. Installation, maintenance,
and test of microwave
equipment _____ _____ _____ _____

V.

a. Did you ever look into a transmission line such as a wave guide
while it was energized? Yes ____ No ____

b. If yes:

How many times?			Average Power?	How Viewed?	
1–3	4–10	over 10		Viewing Bend	Open wave guide
_____	_____	_____	_____	_____	_____
_____	_____	_____	_____	_____	_____
_____	_____	_____	_____	_____	_____
_____	_____	_____	_____	_____	_____

VI.

a. Did you ever feel heat from microwaves coming from a wave guide
or transmission line? Yes ____ No ____

b. If yes, how many times?

	1–3	4–10	over 10
Hands only	_____	_____	_____
Whole body	_____	_____	_____
Head only	_____	_____	_____

c. What types of equipment were involved?

Average Power	_____	_____	_____
Frequency or Band	_____	_____	_____

VII.

a. Have you ever worked near an antenna while it was radiating?
Yes ____ No ____

b. If yes, fill in the following:

	1–3	4–10	over 10	Average Power	Freq. or Band	Distance from Radiating Surface
1. Front Surface						
few seconds	———	———	———	———	———	———
few minutes	———	———	———	———	———	———
over an hour	———	———	———	———	———	———
2. Rear Surface						
few seconds	———	———	———	———	———	———
few minutes	———	———	———	———	———	———
over an hour	———	———	———	———	———	———
3. Sides						
few seconds	———	———	———	———	———	———
few minutes	———	———	———	———	———	———
over an hour	———	———	———	———	———	———

4. How many times did you feel heat from microwaves coming from the antenna?

few seconds   ——— ——— ———

few minutes   ——— ——— ———

over an hour   ——— ——— ———

A Review of Unanswered Biological Hazard Operational Problems

JOHN E. BOYSEN, USAF, MC
Hq Air Materiel Command
Wright-Patterson AFB, Ohio

WE, in the Surgeon's office of Headquarters, AMC, are in the middle concerning matters of occupational medicine. Our problems, including those of microwave radiation, are those of the entire occupational medicine field. One of our responsibilities is to translate research data and information gathered during industrial developmental processes into a form that may be used by the military or civilian worker in the Air Force industrial and operational environment to protect him from his environment. These technical orders or operating instructions must be readily understood by personnel at all technical levels. The end result of our actions must be to either eliminate the hazards or to protect the individual worker so that he may not be injured by any of the hazards in his environment. The same information which we receive from industrial research personnel must also be translated into technical medical information so that the industrial hygiene personnel are given sufficient technical data to enable them to perform their functions in analyzing the working environment and describing the exposure, and so that the occupational medicine physicians may conduct periodic medical examinations or laboratory tests which will enable them to evaluate such exposures and to initiate therapy if indicated. Essentially, then, our mission lies between the Air Research and Development Command and its contractors, on one hand, and the using commands, our "customers," on the other.

Essentially, microwave exposure is an industrial or occupational medicine exposure problem whose solution must follow the same pattern as used in the investigation of the hazards of chemical or other physical agents found in an industrial environment. The solution must depend upon a level of knowledge which includes the following basic factors:

1. Nature of the agent.

2. The manner in which it is absorbed, detoxified, or otherwise rendered impotent, and lastly, dissipated or excreted as the case may be. A description of these phenomena must be capable of being described quantitatively for it is this factor which spells the difference between health, illness, and death.

3. The physio-pathological changes which may be produced by the agent, whether these changes are clinical or subclinical.

4. The means by which the agent may be detected or measured in the environment or in the body fluids of the exposed employee.

5. A thorough knowledge of the control measures which is, in part, based upon the above.

I will try to describe each of these major factors and will bring out certain points in each of these general areas in which I feel there are still unanswered questions.

Nature of Microwaves

The entire electromagnetic spectrum is a continuum from one end to the other (Fig. 1). The various portions of it are different only in the frequency or wavelength. On this chart I have drawn, without any attempt at accuracy, certain effects which may be produced by different radiations. These effects may be physiological or may be pathological. You will notice that the effect in the visual light region is one of sight. In the microwave and infrared bands, thermal effects are noted. You will also see that no ionization is produced by radiations below the ultraviolet level. These effects are all smooth curves. There are no sharp cut-offs. All the effects detected thus far are closely related and vary only slightly in certain aspects from each other.

One other parameter in an electromagnetic field of microwave frequencies is that of power density. In the visible spectrum this parameter is measured in foot candles. In ionizing radiations, it is comparable to the unit of the roentgen. Electronic engineers have been able to describe very accurately, in mathematical terms, the wave patterns from microwave transmitters, whether they emanate

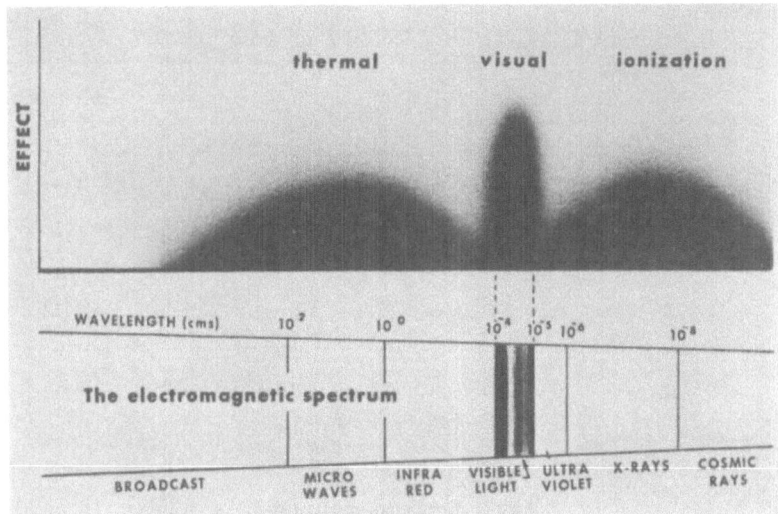

FIG. 1. Effects of electromagnetic radiation.

from wave guides or antennae. They know and understand pretty well the variations of energy within the near and the far zone of the microwave antenna. These all describe the "nature of the agent." I suppose that there are many unanswered questions in this descriptive area, although I don't think that this is one of the major problems at the moment.

Absorption and Dissipation of the Microwave Energy

As far as I have been able to determine, there are still many details in this general area which remain unanswered. For example, I think that it would be important for us to have more information on the quantitative nature of the absorption at the various frequencies (Fig. 2). What is the relationship between the incident energy and that portion which is reflected, that which is transmitted, and the balance of which is absorbed and converted into other forms of energy? We all have a general feeling for this; that is, long wavelengths of the broadcast band are probably very largely transmitted, very little is absorbed, and very little is reflected. In other words, the human body is a fairly transparent object to long wavelengths. However, this is not the case with microwave radiation. Here, there

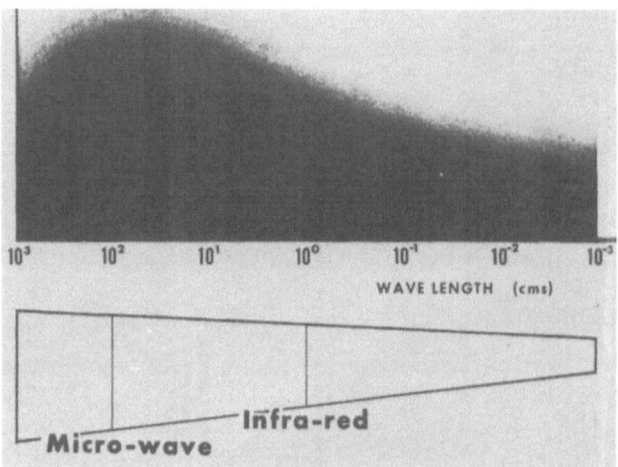

FIG. 2. Coefficient of absorption.

is a considerable amount of reflection. Anyone who has tried to tune in his television set while standing in front of rabbit ear antenna certainly is aware of this problem. The question I am asking is: What are the coefficients of absorption at the various frequencies in which we are interested?

We have heard a great deal about dosage, exposure rates, and power densities. We have been given a maximum permissible dosage level of 10 mw/cm². I seriously question not the figure but some of the approach. In Figure 3, you will notice a family of exponential curves which, in themselves, have no actual validity but would express a relationship between incident energy or power density and the time of exposure. This relationship is of importance. It follows principles which have been worked out in the chemical fields and also have arisen in the field of radiobiology. I see no reason why the same importance should not be attached to this field. As you will notice, I have indicated a line at 10 mw/cm² on the curves, but I would like you to put in the time factors on the coordinate. I am not sure whether the line marking "x-time" is actually measured in hours or minutes. If this level in relationship to the maximum permissible limit is measured in hours, the implications in terms of work time for a radar technician is very significant. On the other hand, if it is only in minutes, then the problem is quite different. In operations throughout the Air Force, this is one of the most pressing problems for which we must have better and more definitive answers.

Another factor about which we should know more is the recovery rate related to frequency and power densities. This is particularly important when we are dealing with a rotating antenna with a rate of revolution of perhaps 6 or 12 or some other interval per minute. A man standing in one place will be radiated for relatively short periods of time. If the recovery period for that frequency and power density is less than the time of onset of the next exposure, there is no problem. His ability to work 8 hr in this environment would be unimpaired regardless of what the power density might be at a certain period of time.

From all the evidence which has been collected to date, I believe it is the majority opinion that the effects of microwave radiation on the human body are thermal in nature. However, I recognize that there are some investigators who have evidence that may indicate that there are or may be nonthermal effects. One suggestion is that such radiation may affect the electrical potentials of the nerve fibers of the central or the peripheral nervous systems. This may or may not be true. I am reminded of the fact that people are occasionally electrocuted, whether accidentally or intentionally, by a form of electromagnetic radiation, although it may be of a different frequency and perhaps with a different power density. This death, I know, is not entirely a thermal one.

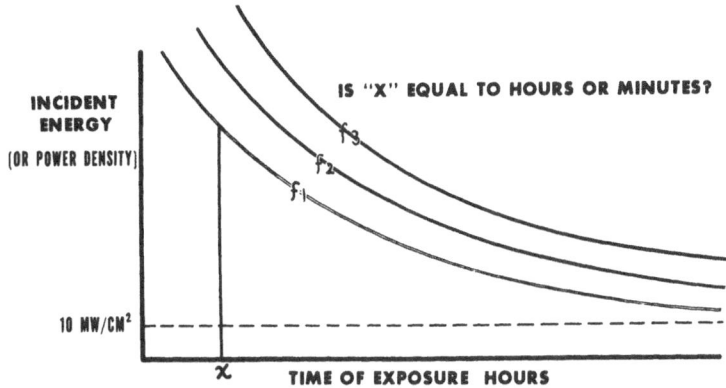

Fig. 3. Dose—time relationship (for various frequencies).

Physio-pathological Changes

What physiological or pathological changes may occur as a result of exposure to microwave radiation? Under some circumstances whole body irradiation will be encountered; under other circumstances, limited regions will be affected. I would like to know which of the organs are at greater risk. With such information, then, we will be better able to establish periodic physical examination procedures which would uncover changes in the most sensitive organ system that would be affected, perhaps before any other organ systems demonstrate any effect whatsoever. It seems, from reviewing the literature, that there is a predominance of information concerning the eyes and testes. I am aware of two things: a great deal of research work has been done on the effect of microwaves on the eyes and on the testicles. The latter always seems to attract initial attention with the advent of any new physical or chemical agent, but I am not sure that this is necessarily logical. It may only be psychological, or due to the fact that these organs are relatively easy to work on. I do recognize the relatively poor heat dissipating mechanisms in the eye. The question is: Which organs are at the greatest risk under the conditions of the *usual* exposures? What pathological changes occur at the various power densities or levels of absorption? Which changes are reversible and which are irreversible? We have heard a great deal about the increased body temperature and increased temperature of the various organ systems. What physiological or chemical changes may occur? I do not believe that a great deal has been done along this line. I know that such changes may be subtle but may be very important. Are there any alterations in enzyme systems? I have already mentioned electrical potential changes as another question.

The Microwave Syndrome

What subclinical signs of overexposure are noted? I am sure that we are all aware of the sense of warmth that a person experiences during exposure and that this is a rather good indication of exposure since it is almost immediate. This must occur before any significant damage can occur unless we assume that certain electrical changes or some obscure chemical changes could occur immediately. Not that I condone such practices, but I am sure that you are all aware of the fact that many technicians, for many years in the past, have made it their practice to warm themselves in the beam of transmitting antennae when the immediate environment was uncomfortably cool. We have not been able to discover, to the best of my knowledge, any damage which has occurred as a result of this practice. This would make one wonder whether it would be likely that an individual would suffer, for example, the formation of lenticular cataracts under the circumstances of ordinary exposures without being forewarned by the warmth that he would sense in the skin about the face or other parts of the body. The same principles would appear to me to apply in the case of other pathological changes; whether these changes were hemorrhagic disorders of the tissues or of internal portions of hollow viscera. It seems to me that such irreversible damage would occur only in the case of those individuals who could not escape from the radiation field even though they did have a forewarning in the form of this sense of warmth.

To complete this particular section, I have also included this question: What is the actual clinical syndrome which one would expect in whole body irradiation of sufficient intensity to produce such a syndrome? Would it involve primarily the eyes, or would he receive skin burns on the side of the body which was exposed? How would it vary with the different frequencies assuming the same power densities? Certainly, it would seem reasonable to expect more superficial burns with a shorter wavelength than one would expect with long wavelengths, particularly those in the broadcast bands to which the body is primarily or essentially transparent. What is so unusual about this? I am sure you have all seen a roast beef prepared in a radar oven. This is one reason why I feel that any maximum permissible level of exposure must be described completely in terms of its frequency, its power density, *and the time of exposure.*

Detection and Measurement

As touched upon earlier in this talk, detection instruments as well as quantitative measurement instruments have been developed which apparently are proving fairly satisfactory. I am sure that those who have been working on these problems will come up with some reasonably accurate devices which will be improved as time goes along, and which will serve the purposes for which they are intended; that is, in measuring the environmental conditions surrounding a microwave generator.

There is one requirement that needs further study and improvement, and that is personnel dosimetry. We must be able to describe in a simple manner the absorbed energy in terms of human tissue. This is a problem which, I know, has plagued the experts in the field of ionizing radiation in much the same manner.

Control Measures

We in the business of occupational health are accustomed to thinking of control measures as applying first, to the source; second, to the environment or the transmitting vector, as it were; and third, to the susceptibility of the individual. Shielding of the source in the case of microwave radiation equipment is virtually impossible because to shield the antenna would merely destroy the use for which the microwave beam was designed. Obviously, it would be safe but it would be of no use to anyone! Modifications in the mode of operation have been accomplished. One of the big problems, however, is the matter of "outage," which means that the power is turned off for a period of repair or adjustment, or it may be automatically turned off while the antenna is "searching" a certain area which is inhabited. Interruption of the beam in this manner is also thought to reduce the operating life of the electronic mechanisms involved.

Control measures as applied to the transmitting vector—the space between the antenna and the man under exposure—may be accomplished by the erection of a shielded work area. This, in many instances, is practical where the work area is by its very nature very small; however, this has its limits as you are well aware.

The last control measure to be considered is that of the susceptibility of the individual who is being exposed. In certain circumstances, I believe this will require the development of personal protective equipment. This problem is being worked on, I know. Some protective suits, for example, have been developed which are effective. However, these are usually cumbersome and interfere with the efficiency of the workman. I do think we need more work in this area.

The last matter concerns medical surveillance. We do need information concerning the procedures which should be incorporated into periodic physical examinations in order to determine any changes which have occurred as a result of exposure, whether these changes are subclinical or not.   I would like those measures or procedures which will detect at a very low absorption level the subclinical, pathological, or physiological changes which have been produced by microwave radiation.

Conclusion

I have tried to suggest some of the unanswered problems concerning the nature of microwave radiation, the manner in which it is absorbed and releases its energy, and the manner in which it is dissipated.   I have tried to present some of the unanswered questions concerning physiopathological changes which may be produced— some of the problems concerning detection and measurement of microwave energies, and lastly, the control measures which are required in the field.

ACKNOWLEDGMENT

I am deeply grateful for the able technical assistance and advice given me in the preparation of this paper by Lt Col Charles C. Dills, Lt Col Robert E. Shirley, and Captain John C. Villforth of the Surgeon's staff, Hq AMC.

Similarities and Differences between the Technical Aspects of the Navy HERO Program for Ordnance and the Personnel Hazard Program

JAMES N. PAYNE
US Naval Weapons Laboratory
Dahlgren, Virginia

THE JOB of the Electromagnetic Hazards Division at the Naval Weapons Laboratory is to direct the technical aspects of the Hazards of Electromagnetic Radiation to Ordnance (HERO) program. The burden of this program is to find ways of eliminating the danger of premature firing of rockets or missiles — or the explosion of their warheads — by stray radio or radar waves. Navy radios and radars are now so powerful, and the rockets and missiles are now so sensitive, that it is, in some circumstances, dangerous to turn on our transmitters when rockets and missiles are nearby. If our radar eyes are shut and our radio voices stilled, the capability of the fleet is seriously reduced. This is an intolerable situation and a solution to the problem is urgently needed.

My paper will be more concerned with HERO than with the personnel hazards program for two reasons: (1) we have to know something about the HERO program before we can make comparisons, and (2) I have first-hand information on the HERO work, but my information on the personnel hazards work has all been obtained from the literature.

Administration

The administration of the two programs is different. The personnel hazards program is primarily under the Air Force and the ordnance program is primarily under the Navy, although, actually, all services are working in both fields. This is a logical approach. A modern land-based air strip is at least 10,000 feet long, so the minimum size of an Air Force base with two crossed runways is $(10,000 \text{ ft})^2$ or 10^8 ft^2. This is approximately 2500 acres. The largest Navy carrier, on the other hand, is of the order of 1000 by 300 ft, or $3 \times 10^5 \text{ ft}^2$. This is approximately 7.5 acres, or a ratio of the order of 300 to 1 in area. The Air Force can separate their ordnance and transmitters by relatively large distance, the Navy cannot. The Air Force cannot separate people and transmitters.

The Air Force undertook the personnel hazards program early and made their findings known to the Navy and Army; therefore, the Navy and the Army did not have to conduct as much research work in this field. The Navy's program is directed at her main problem, ordnance, and her findings in this field are made known to the other services. The application of the research results to particular ordnance and to a particular personnel hazard is the responsibility of the cognizant service.

The HERO work has been underway for less than 2 years. In this period we have determined that a number of environments were hazardous to particular weapons in particular configurations but, at the same time, a larger number of nonhazardous conditions have been discovered. Most of this work has been of the fire drill variety; i.e., someone thinks a weapon may be susceptible in some particular situation and we have to tell him tomorrow whether it is. These are the growing pains experienced by many programs and we expect to have our work in this area on a more systematic basis in the not too distant future.

Frequency

How does frequency affect the hazard? How does it differ for ordnance and for personnel? Our work has shown that in certain weapons it takes less intense rf fields to actuate electroexplosive devices at some frequencies than at others. For example, in a recent test of a weapon at 8, 9, and 10 mc, it was found that the energy available to the electroexplosive device (EED) was 5 times larger at 9 mc than it was at either 8 or 10 mc. The point here is that the weapon system, acting as an electrical transmission line to the electroexplosive device, has frequency discriminating characteristics.

Let's imagine a plot of hazard as a function of power level, duration of exposure, and frequency. We'll not be too precise about our definition of hazard but a reasonable definition might be the probability of actuation multiplied by a weighting factor that takes into account the seriousness of an unintentional firing. For example, on any sort of reasonable scale, a probability of firing of .01 would be a much more hazardous situation for a nuclear warhead than, say, for a flare or flash signal. The same notion could apply to the personnel hazards problem. A cataract br a permanent injury to the testes would be a greater hazard than superficial burning or a moderate whole-body temperature rise.

Now what does our hazards plot look like? I think it is safe to say that both ordnance and personnel hazards depend on frequency. Our evidence indicates that for ordnance the hazards are greater in the communication bands than in the radar bands. The papers at this meeting, and previous work, indicate that the personnel hazards are greatest in the L to S radar regions of the spectrum although your restriction of 10 mw/cm^2 is stated without respect to frequency.

Measuring Methods

In determining the hazard to ordnance we measure the rf energy dissipated in the small electric heater element of the EED. The most common variety of EED contains a small electric heater surrounded by a temperature-sensitive explosive. Normally this device is operated by supplying electrical energy to it under controlled conditions. Of course our hazard is generated when rf energy gets into the electric heater inadvertantly. This device does not know whether it is rf or dc. If energy is dissipated in the electric heater, the explosive gets hot and an explosive train is actuated.

In determining the hazard we use EED simulators. The EED simulator is a standard EED with all of its explosive removed. A thermocouple is mounted within the EED canister in place of the explosive. The thermocouple senses the temperature rise in the EED electric heater and, knowing the temperature required to actuate the device, the hazard is determined for that environment and the particular weapon system configuration. The output of the simulator is fed to a recorder. The complete measuring system is contained within the electromagnetic shield of the weapon to keep the perturbations to a minimum. The system I have outlined is the most generally used technique, but thermistors and bolometers are also used. The measuring method I have described determines the hazard for a particular test condition. That is, in a single test we get data for only one particular set of values for the many variables that characterize the problem. These tests are time consuming and expensive, and therefore the hazard has to be determined from a comparatively few tests. One of our more difficult problems has been to determine the conditions under which a weapon system is most hazardous. Of course we will not live long enough to test any weapon under all conditions, and if we don't happen to hit the worst condition fortuitously somebody else may not live long enough to tell us we missed it. Therefore, the necessity for predicting the most hazardous conditions is paramount.

I have not mentioned it, but go/no-go tests are conducted. In these tests standard EED's are used instead of EED simulators. If an actuation does not occur, very little information is obtained.

The testing method outlined above has serious limitations and the HERO group has been seeking a more general method. The method outlined above is limited by the requirement that a most hazardous condition be explored in a few exposures, say less than 10, out of an infinite number of possible conditions. Also it requires testing in large rf fields. The Chief of Naval Operations has been very generous in assigning frequencies for this work; even so, we can radiate at high power on only a relatively small number of frequencies. You can see we needed a better system, and we are presently very hopeful for a system devised by the Jansky and Bailey Co. in cooperation with Mr. Potter of our Evaluation Branch. The Jansky and Bailey system will allow one to characterize the electrical system, excluding the weapon, at low energy level as a function of frequency; and to independently measure the electrical characteristics of the ordnance in the laboratory. The hazard to the system as a function of frequency and rf power density can then be predicted by properly combining these results.

The literature with which I am familiar does not discuss instrumentation for personnel hazards in detail. You do use thermocouples, thermistors, thermometers, etc., and you have essentially the same requirements for field strength and power density measurements as those required for the HERO program. I think that we can say that both programs need the same kind of instrumentation; i.e., equipment to measure the environment and thermometry to measure the temperature rise in a specimen exposed to the environment.

Restrictions

Restrictions have been imposed upon the Army, Navy, and Air Force because of radio frequency hazards. This is a serious situation. Every restriction results in a reduced capability of our fighting forces, and therefore fleet commanders oppose the restrictions. They emphasize we cannot afford the restrictions and they ask for safe ordnance. A restriction in the case of ordnance denies the ship the use of its radio and/or radars when handling ordnance. Communications is a problem of great magnitude without this nasty little gimmick thrown in. Restrictions resulting from personnel hazards also limit fleet capability.

The personnel hazards seem to be most pronounced at the radar frequencies. It is for these frequencies that we have extremely high gain antennas that give very large field intensities. The restriction of a maximum power density of 10 mw/cm² has caused some inconvenience for these highly directional arrays. At lower frequencies, e.g., in the communications bands, this restriction has not really been a restriction because power densities of this magnitude are not encountered. Thus a single limiting power density restriction, stated without respect to frequency, has been blanketed across the spectrum without serious outcry. However, this happy situation is likely to change. When we begin to use transmitters in the 1 to 30 mc frequency range which will produce power densities greater than 10 mw/cm², I believe that a great deal of pressure will be placed upon you to determine the hazard as a function of frequency. Transmitting equipment capable of producing these power densities is just around the corner.

Ordnance restrictions are specified for particular weapons in specific environments. An ordnance component will act as an antenna system with a possible net gain, particularly at frequencies for which the system elements are resonant. As we identify the hazardous frequencies and the dangerous power levels, restrictions are recommended. I think you can see that the fleet would have been out of business if the minimum field required for actuating the most sensitive ordnance device had been imposed as a restriction upon the fleet. It would have resulted in complete transmitter silence anytime ordnance was being handled or used. Restrictions have only been imposed upon ordnance when tests on the particular ordnance item showed it to be unsafe. We know there is a large gray area between safe and unsafe, and we in the Navy working on the ordnance problem are gambling on minimum restriction. The personnel hazards people have imposed the maximum restriction. Your maximum restriction has not seriously interfered with operations; our minimum restrictions have interfered. As we learn more about our respective problem, I expect to see a relaxing of restrictions upon personnel at both ends of the radio frequency spectrum and an increase in restrictions imposed on ordnance until improved system designs are in service.

Protective Devices

The restrictions we were discussing above are necessarily interim measures. The ordnance program has to develop rf-safe ordnance, and the personnel hazards program has to either evolve rf-safe people or protect the unsafe people with protective clothing.

A relatively large portion of our work is directed toward providing rf-proof ordnance. We are approaching this problem from two directions. The first is an attempt to provide ordnance which will reject all electrical excitation in the frequency range, from say, 10,000 cps and higher, but have the same response to dc and low frequency signals as the present electroexplosive devices. This I consider the ideal solution since ordnance design engineers would continue working with the kinds of devices with which they are familiar and aircraft and ship firing circuits would not have to be modified. In case this utopia is not realized, we are concurrently conducting research on new concepts, such as exploding bridge wire initiators and conductive film igniters. These devices would require new firing systems and a reorientation of well-established practices by ordnance designers.

My information on protective clothing for personnel hazards is that reflective devices only are being considered. This seems reasonable, since you may be able to keep rf from the person by using clothing which would absorb the rf energy, but which might cook him in large fields as the clothing heats up.

Conclusion

I think we can say that the two areas of hazards work are basically similar. Both hazards result from an electromagnetic environment and in both cases the hazard is principally thermal; i.e., rf energy is dissipated in a load which creates a temperature increase. So the differences are a matter of degree. Of course there is a fundamental difference between the molecules of which a human being is constructed and the molecules in a missile, but the same physical parameters can be used to describe both in dealing with the radio frequency problem. The amount of energy required to produce a given hazard is a function of frequency for humans and for ordnance. The same kinds of instrumentation are used in determining the hazard for both problems. Restrictions are different, but only in degree. There are more fundamentally different kinds of protective devices available to the ordnance designer than to the personnel hazards worker, but the reflective technique used for personnel is also effective for ordnance.

APPENDIX

The paper listed below was presented as a part of the conference but was not available for inclusion in these Proceedings.

Naval Exposure Environment

LT. COMMANDER F. E. EDMUNDS
Bureau of Ships
Washington 25, D.C.

INDEX

Absorption,
  cross sections, 153, 172
  electromagnetic energy, 117
  function of body size, 164
  function of conductivity, 163
  microwave energy, 150, 153, 202, 215, 221, 239
Accessory sex glands, 99
ADDINGTON, C. H., 177
Ambient temperature, 279, 283
ANDERSON, W. A. D., 99
Androgen, 99, 194, 199
Anechoic chamber, 164, 174, 202
ANNE, A., 153
Antennas, 4, 6, 16, 24, 53, 56, 164
  corner reflector, 201
  distance, 58
  effect, 179
  illumination, 61
  round, 61
  scanning, 66
  square, 64
Antigenic reactivity, 118, 132
Aqueous humor [see also: Eyes], 290
Ascorbic acid, 285
Athermal effects [see: Nonthermal effects]
Autopsies, 183, 234

BACH, SVEN A., 117

BAUS, R., JR., 229
BERGER, CAROLYN, 251
Bio-negative action, 222
Bio-positive action, 222
Blood circulation,
  dogs, 190, 269
  eggs, 216
Body weight, 179, 272, 278
Boltzmann distribution, 86, 138
BOYSEN, JOHN E., 309
Brain, 232
BROWNELL, ARNOLD S., 117
Brownian motion, 138, 224

Cataractogenic effect, 201, 285, 293
Cats, 232, 251
Cellular effects, 135, 217, 226, 265
Chemical effects, 225
Chronic irradiation, 135, 177
Collimated beam, 66
Colloidal systems, 117
Cranial effects, 188, 231, 252
Crossover region, 50

DAHLEN, ROGER W., 187
Debye effect, 224
Debye's equation, 125, 127
DEMINCO, ANTHONY P., 33
Dogs, 143, 177, 178, 187, 201, 230, 261, 263

Dosage, 95, 179, 182, 222, 225
  critical, 179
Dosimeter, film badge, 43

Edema, 270
EDMUNDS, F. E., 327
EISENBUD, MERRIL, 293
Electric constants,
  biological liquids, 222
  body tissues, 155
Electrical effects, 224
Electrophoretic patterns, 118
  controls, 124
  double peaks, 123, 128
  harmonics, 125, 126
  relative humidity, 127
  temperature, 123, 126, 129
Electrostatic interaction, 137
Embryo, chick, 201
Endocrine chain (pituitary—
  testes—prostate), 101, 108
ENGELBRECHT, R. S., 55
Environment, radio-frequency, 3,
  71
Exposure, infrared [see: Infrared
  radiation]
Exposure, microwaves,
  ambient temperature, 279, 283
  brain, 233, 270
  index, 300
  long-term, 136
  maximum safe level, 6, 9, 15,
    47, 144, 283, 312
  neural tissues, 258
  pituitary, 110
  power levels, 5, 31
  scrotum, 105, 111
  survival, 179, 182, 216
  testicular damage, 105, 192
  time factor, 30, 205, 283

Zn^{65} uptake by dorsolateral
  prostate, 105
Exposure, X-ray [see: X-ray expo-
  sure]
Eyes, 15, 78, 201, 216, 285, 293

Far field, 20, 24, 48, 56, 172
Field strength [see also: Power
  density], 57, 96, 105, 120,
  179, 181, 190, 198, 226, 263,
  271
  locating maxima, 16
  measuring equipment, 16
  measurements, 15, 23, 152, 202
  pearl-chain formation, 90
FISCHER, F. P., 177
FLEMING, J., Jr., 229
Fraunhofer region, 56
Frequency spectrum, 34, 311
Fresnel region, 56, 149
Fruit flies, 187, 195

Gamma globulin, 117
Gibbs ensemble, 138
Glutathione, 287
Gonadotrophin, 99
GOULD, THELMA CLARK, 99
Guinea pigs, 177, 182
GUNN, SAMUEL A., 99

Hematocrit values, 264, 272, 274,
  278, 280
Hemodilution, 276
HOWLAND, JOE W., 261
Human gamma globulin, 117,
  132
Human subjects, 232, 293
Humidity, relative, 127
Hydration, 274

Illumination, tapered, 61
IMIG, CHARLES J., 187
Industrial hygiene, 13, 309
Infrared radiation, 112, 187, 192, 195, 239, 253
Instrumentation [see also: Measuring equipment], 16, 18, 22, 23, 28, 41, 73, 111, 119, 140, 141, 145, 146, 167, 184, 242, 244
Intensification factors, 226
Intermittent exposure, 279
Intermolecular activity, 138
Interstitial cells in testes, radioresistance of, 102
Ionizing radiation [see also: X-ray], 281, 287, 299
Irradiation,
  chronic, 135, 177
  specific area, 27
  whole body, 26
Isolated cells, chick osteoblasts, 216

KINOSHITA, JIN H., 285
Kirkwood—Shumaker forces, 138
Klystrons, 36
  oscillating, 40
KNAUF, GEORGE M., 9

Lens of eye, 201, 203, 285, 293
  chemical composition, 286
  defects in humans, 293
  luminescence, 296
  minute defects, 296
  normal characteristics, 294
  opacification, 201, 295, 296
  posterior polar defects, 297
  sutural defects, 297
Lenticular changes, 301
Leucocyte changes, 263

London — Eisenschitz — Wang forces, 138
London—van der Waals' dispersion force, 137
Longevity, 135
LUZZIO, ANTHONY J., 117

Magnetron, 37
Male endocrine system, 99, 192
Maximum safe exposure level, 6, 9, 15, 45, 47, 144, 283, 312
  international acceptance of, 10
Maxwell's equations, 154
MCAFEE, R. D., 229, 251
MEAHL, HARRY R., 15
Measuring equipment, 16, 23, 150
MERMAGEN, HERBERT, 143
MEROLA, LORENZO O., 285
Mice, 135, 230
MICHAELSON, SOL M., 261
Microwave frequencies,
  spectrum, 34, 311
  0.1–1000 mc, 74
  8–10 mc, 320
  10–200 mc, 123
  100–35,000 mc, 239
  200 mc, 177, 215, 261
  300 mc, 174, 225
  400 mc, 163
  500 mc, 175
  1000 mc, 17, 174, 225
  1000–10,000 mc, 150
  2450 mc, 187, 201, 285
  2700 mc, 30
  2800 mc, 261
  2880 mc, 159
  3000 mc (10 cm), 139, 143, 174, 230
  9000 mc, 18

Microwave frequencies (*cont.*)
  10,000 mc (3 cm), 163, 230,
    251
  24,000 mc, 99
Mitotic cells, 266
Molecular response, 139, 226
MORESSI, WILLIAM J., 187
MUMFORD, W. W., 55

Naval applications, 47, 319, 327
Near field, 6, 24, 49, 56, 172
NELSON, NORTON, 1
NEUBAUÈR, R. A., 177
Neurological effects, 229, 251,
  258
Nociceptive response, 229, 251
  definition, 238
Nonthermal effects, 113, 177,
  217, 222, 226, 252

OSBORN, C., 177
OVERMAN, H. S., 47

PAYNE, JAMES N., 319
Pearl-chain formation, 85, 224
  biological significance, 96
  thresholds, 95
Penetration of tissues, 239
Personnel, protection of, 5, 16,
  21, 45, 68, 71, 293
Perturbation theory, 140
Phantoms, 143, 153, 154
  field patterns, 144
PINNEO, L., 229
Pituitary gland, 100
PIZZOLATO, PHILIP, 251
Plane waves, 155
Polarity (of rf field), 178
Power density [*see also:* Field
  strength], 24, 132, 135, 169,
  178, 239

aperture, 24, 61
distance, 51, 60
calculation of, 56
exposure time, 30
radar beam, 48
Power levels in exposure, 5, 6, 31
PRAUSNITZ, SUSAN, 135
Prostate gland, 100
Protection of personnel, 21
barriers, 46, 67
devices, 324
interlocks, 45, 68
monitoring, 16, 45
protective garment, 71
sector blanking, 5
shielding, 38, 40, 68
Pulsed microwave energy, 30, 94,
  226, 261
Pulsed X-rays, 33

Rabbits, 187, 201, 216, 285
Radar safe distances, 47
Radar, sites, 4, 47, 320
Radiation hazards, 55, 309, 319
Radiometer, 242
Rapid-flow apparatus, 141
Rats, 99, 143, 187, 192, 231, 264
Reflection phenomena, 146, 240,
  312
Resonator, cavity, 140
REYNOLDS, MARTIN R., 71

Safe distances,
  formulas, 48
  radar, 47
SAITO, M., 85, 153
SALATI, O. M., 153
SARKEES, Y. T., 177
Scanning antennas, 66
Scattering theory, 155
SCHWAN, H. P., 85, 153

SEARLE, GORDON W., 187
Sensory motor cortex, 238
Shielding, 20, 29, 71
  lead, 40
  leaded glass, 38
  metalized cloth, 72
  metallic screen, 68, 78
  steel, 38
Smoluchowski equation, 88
Specificity effects of molecular
  processes, 137
Spermatogonia, 102
Square antenna, 64
Surgical techniques, 236
Survival time, 190
SÜSSKIND, CHARLES, 135
SWARTZ, G., 177
Symptoms, exposed dogs, 180

TALLMAN, OLIVER G., 3
Tapered illumination, 61
Telemeters, 185
Temperature measurement, 10,
  29, 111, 127, 135, 143, 149,
  170, 178, 183, 189, 195, 202
Testicular effects, 105, 110, 192,
  197
Testosterone, 110, 194
Thermal effects, 95, 105, 170,
  180, 214, 216, 226, 252, 257
Thermal stress, 267, 270, 276
THOMAS, JOHN A., 187
THOMSON, JOHN D., 187
THOMSON, RODERICK A. E., 261
Thyratron, 38
Time factor [see also: Dosage], 30,
  96, 143, 222, 283
TOMBERG, VICTOR T., 221
Transmitter power, 61
Traveling wave tube, 36

VAN UMMERSEN, CLAIRE, 201
Visual acuity, 301
Visual cortex, 231
VOGELHUT, PAUL O., 135
VOGELMAN, JOSEPH H., 23

Waveguides, 27
Wavelength in tissues, 240
Whole body radiation, experi-
  mental, 26, 102, 154
  partial exposure, 27, 163, 257,
  261
Whole brain polarization, 231
WUNDER, CHARLES C., 187

X-ray detection, 41
  ionization chambers, 41
  Geiger tubes, 41
  photographic dosimetry, 43
  scintillation-type detectors, 43
X-ray exposure, 3, 12, 44, 217
  hard, 35
  human gamma globulin, 117
  male endocrine system, 102
  pituitary, 104
  pulsed, 33
  rabbit eyes, 216
  soft, 35, 37
  spermatogonia, 102
X-ray sources, 36
  klystrons, 36, 40
  magnetrons, 37
  microwave equipment, 33
  shielding, 38
  thyratrons, 38
  traveling wave tubes, 36

ZARET, MILTON M., 293
Zn^{65} uptake,
  dorsolateral prostate, 101, 106
  exposure to microwaves, 105
  significance, 100